Lecture Notes in Networks and Systems 1679

The series "Lecture Notes in Networks and Systems" publishes the latest developments in Networks and Systems—quickly, informally and with high quality. Original research reported in proceedings and post-proceedings represents the core of LNNS.

Volumes published in LNNS embrace all aspects and subfields of, as well as new challenges in, Networks and Systems.

The series contains proceedings and edited volumes in systems and networks, spanning the areas of Cyber-Physical Systems, Autonomous Systems, Sensor Networks, Control Systems, Energy Systems, Automotive Systems, Biological Systems, Vehicular Networking and Connected Vehicles, Aerospace Systems, Automation, Manufacturing, Smart Grids, Nonlinear Systems, Power Systems, Robotics, Social Systems, Economic Systems and other. Of particular value to both the contributors and the readership are the short publication timeframe and the world-wide distribution and exposure which enable both a wide and rapid dissemination of research output.

The series covers the theory, applications, and perspectives on the state of the art and future developments relevant to systems and networks, decision making, control, complex processes and related areas, as embedded in the fields of interdisciplinary and applied sciences, engineering, computer science, physics, economics, social, and life sciences, as well as the paradigms and methodologies behind them.

Indexed by SCOPUS, EI Compendex, INSPEC, WTI Frankfurt eG, zbMATH, SCImago.

All books published in the series are submitted for consideration in Web of Science.

For proposals from Asia please contact Aninda Bose (aninda.bose@springer.com).

Teresa Zielinska
Editor

Smart Solutions in Power Engineering, Robotics and Human Centered Systems

Selected Works of the 2025 3rd International Conference on Power and Automation Engineering

Springer

Editor
Teresa Zielinska
Faculty of Power and Aeronautical Engineering (FPAE)
Warsaw University of Technology
Warsaw, Poland

ISSN 2367-3370 ISSN 2367-3389 (electronic)
Lecture Notes in Networks and Systems
ISBN 978-3-032-08220-6 ISBN 978-3-032-08221-3 (eBook)
https://doi.org/10.1007/978-3-032-08221-3

This Springer imprint is published by the registered company Springer Nature Switzerland AG
The registered company address is: Gewerbestrasse 11, 6330 Cham, Switzerland

Preface

We are pleased to present the proceedings of the 2025 3rd International Conference on Power and Automation Engineering (PEAE), which was held in Bangkok, Thailand, from June 5–7, 2025. Co-sponsored by Chulalongkorn University, Thailand, this conference has once again served as a premier platform for professionals, researchers, and scholars in the fields of power engineering and mechanical automation to come together, share their knowledge, and exchange innovative ideas.

In today's rapidly evolving technological landscape, power engineering and automation play pivotal roles in driving global development. Power engineering is the backbone of modern society, ensuring a reliable and efficient supply of energy to meet the ever-increasing demands of various sectors, from industry and transportation to residential and commercial applications. Automation, on the other hand, has revolutionized industrial processes, enhancing productivity, precision, and flexibility while reducing human error and labor costs. The integration of these two fields has led to the development of smart grids, intelligent manufacturing systems, and advanced control technologies, which are essential for achieving sustainable development and competitiveness in the 21st century.

The papers included in this proceedings represent the latest research and developments in power and automation engineering. They cover a wide range of topics, such as power system operation and control, renewable energy integration, energy storage technologies, industrial automation and robotics, artificial intelligence and machine learning in power and automation, and power electronics and drives. Each paper has undergone a rigorous peer-review process to ensure its quality and relevance, and they present novel solutions, theoretical analysis, experimental results, and practical applications that contribute to the advancement of the field.

We would like to express our sincere gratitude to all the authors who submitted their papers to the conference. Their dedication and hard work have made this proceedings possible. We also extend our heartfelt thanks to the members of the program committee, whose expertise and efforts in reviewing the papers have ensured the high quality of the selected contributions. In addition, we are indebted to the sponsors, organizers, and volunteers for their generous support and assistance in making the conference a great success.

We hope that readers will find the papers in this proceedings informative and inspiring, and we look forward to your continued support and participation in future PEAE conferences.

Teresa Zielinska
Proceedings Editor

Committees

Conference Advisory Committee

Makoto Iwasaki (IEEE Fellow)	Nagoya Institute of Technology, Japan
C. W. Lim	City University of Hong Kong, Hong Kong, China

Conference Local Chair

Somkiat Tangjitsitcharoen	Chulalongkorn University, Thailand

Conference Chair

Angsumalin Senjuntichai	Chulalongkorn University, Thailand

Conference Co-chair

Teresa Zielinska	Warsaw University of Technology, Poland

Conference Organizing Chair

Haruetai Lohasiriwat	Chulalongkorn University, Thailand

Conference Program Chairs

Zengtao Chen	University of Alberta, Canada
Young Moon	Syracuse University, USA
Oran Kittithreerapronchai	Chulalongkorn University, Thailand

Conference Publicity Chair

Somchai Puajindanetr	Chulalongkorn University, Thailand

Conference Technical Program Committees

Michal Gdula	Rzeszow University of Technology, Poland
Saket Rathi	Applied Materials Inc., USA
Tumelo Seadira	University of South Africa, South Africa
Madan Lal Chandravanshi	Indian Institute of Technology, India
Shahad Ali Hammood Al-Kawaz	University of Babylon, Iraq
Mandlenkosi George Robert Mahlobo	University of South Africa, South Africa
Hechuan Song	University of Science and Technology Beijing, China
Hoo Tien Nicholas Kuan	Universiti Malaysia Sarawak, Malaysia
Ana Margarida Bento	University of Porto, Portugal
Chin Seong Lim	University of Nottingham Malaysia (UNM), Malaysia
Youtian Zhang	Rice University, USA
Wegdan Wagdy	Horus University, Egypt
Youtian Zhang	Rice University, USA
Amine el Mahdi Safhi	Concordia University, Canada
Mark Jackson	Kansas State University, USA
Wahida Tina	Concordia University, Canada & Lansing Board of Water & Light (BWL), USA
Osman Adiguzel	Firat University, Turkey
Affaf K. Al-Oufy	Galala University, Egypt
Persia Ada Narag de Yro	Mapua University, Philippines
Mohamed G. Nassef	Egypt-Japan University of Science and Technology, Egypt
Montri Sangsuriyun	Nakhon Phanom University, Thailand
Rusdi Nur	Politeknik Negeri Ujung Pandang, Indonesia
Babatunde Abiodun Obadele	Botswana International University of Science and Technology, Botswana
Vladoiu Rodica	Ovidius University of Constanta, Romania
Somashekhar S. Hiremath	Indian Institute of Technology Madras, India
Guelailia Ahmed	Algerian Space Agency, Satellite Development Center, Algeria
Koorosh Gharehbaghi	RMIT University, Australia
Trong-Phuoc Huynh	Can Tho University, Viet Nam
Marina Rynkovskaya	RUDN University, Russia
Rahul Davis	Sam Higginbottom University of Agriculture, Technology & Sciences (SHUATS), India
Ravi Kumar Puli	National Institute of Technology, India
Galina Ilieva	University of Beira Interior, Portugal

Contents

Aircraft Wing Design and Aerodynamic Characteristics Analysis

Nonlinear Dynamic Response and Stability of Quadrotor UAVs Under Transient Impact Loading from Specialized Actuation Systems

Zizhuo Cai[1](✉), Gaoyu Liang[2], Maria Sergeevna Selezneva[2], and Mo Yang[3]

[1] School of Automation Science and Electrical Engineering, Beihang University, 37 Xueyuan Road, Haidian District, Beijing 100191, China
tzuchotsai@buaa.edu.cn

[2] Department of Automatic Control Systems, Bauman Moscow State Technical University, 2nd Baumanskaya Street, 5/1, Moscow 105005, Russia
lyang@student.bmstu.ru, ms.selezneva@bmstu.ru

[3] School of Mechanical Engineering, Nanjing University of Science and Technology, 200 Xiaolingwei Street, Nanjing 210094, China
yangmo105@njust.edu.cn

Abstract. Quadrotor unmanned aerial vehicles (UAVs) have become integral to modern industrial applications, particularly in scenarios requiring robust performance under extreme mechanical loading conditions. This study investigates the dynamic behavior of quadrotor systems when subjected to high-intensity impact loads generated by specialized actuation devices. The research establishes a comprehensive framework for analyzing transient mechanical impacts on UAV platforms, beginning with the characterization of impulse loading profiles derived from experimental measurements. Using an industry-standard quadrotor model, the study examines both axial and oblique impact scenarios through systematic numerical simulations. Key dynamic response parameters including displacement, velocity, and orientation angles are quantitatively evaluated. The simulation results reveal distinct behavioral patterns under different loading configurations, providing fundamental insights into energy dissipation mechanisms during impact events, transient stability characteristics and recovery dynamics of the control system. The findings contribute to the advancement of impact-resistant UAV design, offering practical guidelines for enhancing operational safety in demanding industrial environments. This work establishes a foundation for future research on robust aerial systems capable of withstanding extreme mechanical disturbances.

Keywords: Quadrotor UAV dynamics · Transient impact loading · Nonlinear system response · Computational dynamics modeling · Robustness performance analysis

1 Introduction

Small quadrotor UAVs possess inherent advantages for reconnaissance and tactical missions involving specialized mechanical devices due to their lightweight and stealthy characteristics, high maneuverability, and autonomous capabilities in path planning and

T. Zielinska (Ed.): PEAE 2025, LNNS 1679, pp. 3–15, 2026.
https://doi.org/10.1007/978-3-032-08221-3_1

dynamic flight parameter adjustment. The military relevance of such platforms is well-established, exemplified by the BGM-34A UAV which pioneered military applications in 1970 [1], marking a significant milestone in unmanned combat aerial vehicle (UCAV) development. Research by Cai et al. [2] demonstrated the efficacy of global search algorithms for quadrotor path planning and mission execution in both civilian and military contexts. Investigations into platform-weapon interactions include studies on the muzzle response characteristics of Small Unmanned Ground Vehicles (SUGVs) during small arms fire by Ding et al. [3], and the design and adaptability analysis of UAV firearm unlocking force measurement devices by Yang et al. [4, 5]. Furthermore, Prasad et al. [6] quantified rifle recoil forces on UAVs and implemented mitigation strategies. Collectively, these studies confirm the capability of specialized military devices to perform designated tasks on UAV platforms.

Given the operational profile of such integrated systems, blast-induced shock loads represent a critical factor influencing the dynamic response and flight endurance of small UAVs. This is highlighted by reviews on impact testing methodologies for lightweight UAVs [7] and investigations into collision dynamics, impact loads, damage mechanisms, and the influence of attitude angles [8]. Consequently, the effects of explosive shock loads on small UAVs remain a significant research focus within the field.

This study specifically addresses the high-dynamic, high-overload environment inherent to quadrotor UAVs integrated with specialized explosive devices. We focus on characterizing the blast impact loads generated by a specific military device and analyzing the resulting dynamic response of a small quadrotor UAV. Through system modeling and simulation, we quantify the UAV's displacement, velocity, and attitude response under these transient loads, establishing conclusive findings regarding its behavior.

2 Modeling of Blast Impact Loading of Specialized Mechanical Device on Quadrotor UAV

The coordinate system and attitude angles of the system are delineated as depicted in Figs. 1 and 2. Herein, the Earth-fixed coordinate frame $o_e x_e y_e z_e$ is employed to examine the dynamic states of the multicopter in relation to the Earth's surface and ascertain its three-dimensional position. Concurrently, the aircraft-body coordinate frame $o_b x_b y_b z_b$ is affixed to the multicopter, with the center of gravity serving as the origin o_b of $o_b x_b y_b z_b$.

To facilitate analysis and simulation, the model of the quadrotor and specialized mechanical device is simplified as follows:

- Assumed the quadrotor and specialized mechanical device to be rigid bodies;
- Their mass and moments of inertia are constant;
- The impact force from explosives directly acts on the center of mass.

2.1 Blast Impact Loading of Specialized Mechanical Device

Based on the fundamental principles and assumptions of ballistics [9], the operational characteristics of the specialized mechanical device can be obtained as follows. It is noteworthy that the Euler angles presented here represent the ballistic Euler angles of

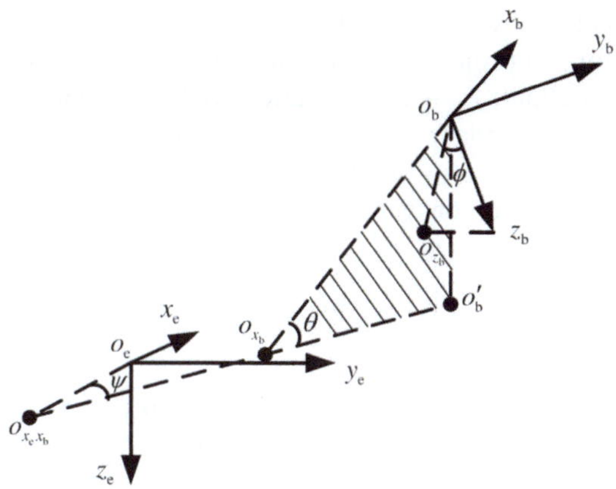

Fig. 1. The Earth-Fixed Coordinate Frame ($o_e x_e y_e z_e$), Aircraft-Body Coordinate Frame ($o_b x_b y_b z_b$) and representation of Euler angles.

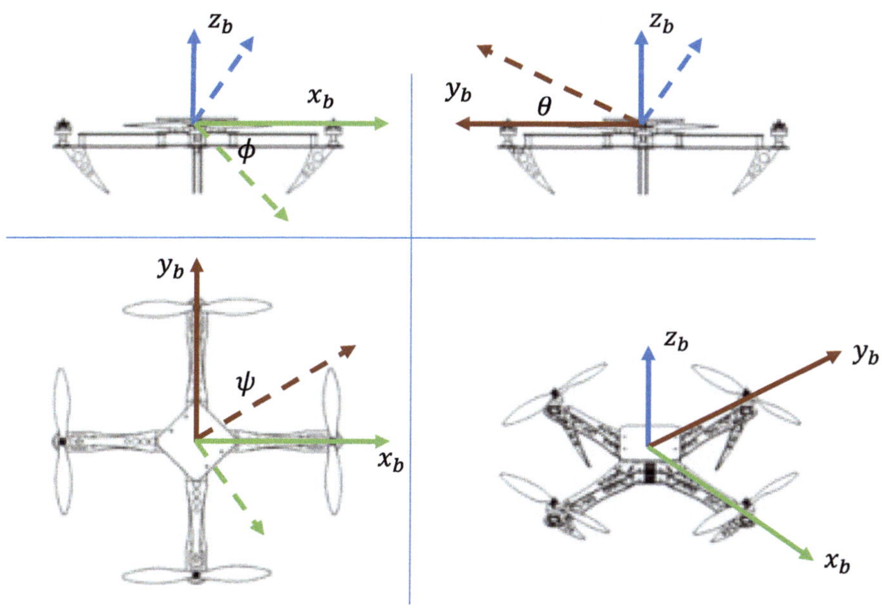

Fig. 2. The intuitive representation of Euler angles, where pitch angle $-\theta$; roll angle $-\phi$; yaw angle $-\psi$.

the specialized mechanical device, denoted as θ_g, ϕ_g and ψ_g, distinct from the Euler angles of the quadrotor aircraft denoted as θ, ϕ and ψ. Readers are advised to take note of this distinction within the text.

The fundamental equations

$$SP\left(l_{\psi_g} + l\right) = f\omega\psi_g - \frac{\theta_g}{2}\varphi_g m v^2 \tag{1}$$

where S — cross-sectional area of the bore; P — average pressure inside the bore; l — travel distance of the projectile within the bore; $\Delta\omega$ — relative weight loss due to combustion of propellant; f — force exerted by the propellant; ω — mass of the propellant charge; η — secondary power coefficient; m — mass of the projectile; v — velocity of the projectile head; k — adiabatic index, the ratio of specific heat at constant pressure c_p to specific heat at constant volume c_w; l_{ψ_g} — free volume of the chamber neck, where

$$l_{\psi_g} = \frac{1}{S}\left[W_0 - \frac{\omega}{\delta}\left(1 - \varphi_g\right) - \alpha\omega\psi_g\right] \tag{2}$$

where W_0 — chamber volume; ρ — propellant density; α — residual gas volume of the propellant.

The motion equations

$$SP = \varphi_g m \frac{\mathrm{d}v}{\mathrm{d}t} \tag{3}$$

According to the geometric combustion law of specialized mechanical device propellants

$$\psi_g = \chi z\left(1 + \lambda z + \mu z^2\right) \tag{4}$$

where χ, λ, μ — propellant shape characteristic parameters.

The proportional burning rate equation

$$\frac{\mathrm{d}z}{\mathrm{d}t} = \frac{P}{I_K} \tag{5}$$

where I_K — total impulse pressure at the moment of propellant combustion completion.

According to the Bravon empirical formula

$$p = p_k e^{-At} \tag{6}$$

where p — average pressure within the bore at a certain instant during the afterburning period; p_0 — average pressure within the bore at the instant of projectile exiting the muzzle; t — time from the beginning of the afterburning period to a certain instant; A — constant coefficient.

Upon computation, the corresponding bore pressure and force variation over time during the internal ballistic phase of this particular mechanical device are depicted in Figs. 3 and 4.

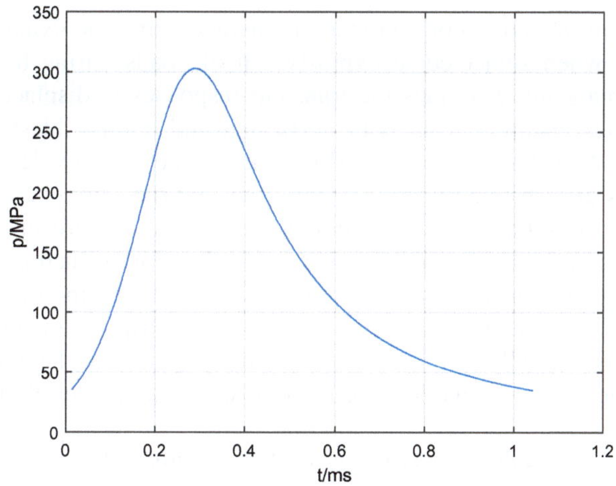

Fig. 3. The bore pressure force variation over time during the internal ballistic phase.

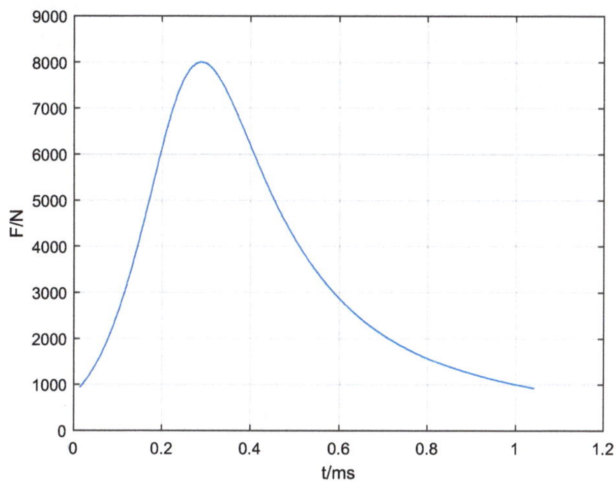

Fig. 4. The bore force variation over time during the internal ballistic phase.

2.2 Quadrotor UAV

The quadrotor UAV platform selected for this research is a certain model of DJI drone, with the key kinematic parameters outlined in Table 1.

The control architecture employed in this study, as illustrated in Fig. 5 and documented in prior research [10–12], utilizes a well-established PID control framework consisting of two interconnected control loops. The inner attitude loop regulates the aircraft's angular orientation, while the outer position loop manages translational movement, collectively ensuring stable flight and precise positioning. Given the algorithm's widespread adoption and thorough validation in existing literature, this manuscript foregoes detailed exposition of its implementation specifics.

Instead, our investigation concentrates on characterizing the dynamic behavior of quadrotor UAVs when subjected to explosive shock loads during hover conditions. The study systematically examines the transient responses in displacement, velocity, and attitude angles resulting from both horizontal and oblique blast impacts generated by specialized mechanical devices. This focused analysis provides critical insights into the aircraft's disturbance rejection capabilities and recovery characteristics under extreme loading conditions, while intentionally leveraging the proven performance of the conventional control architecture as a stable baseline for evaluation.

The research maintains this specific scope to deliver novel findings regarding blast impact dynamics without duplicating well-documented aspects of quadrotor control systems. This approach allows for concentrated investigation of the aircraft's physical responses to explosive forces while building upon established control methodologies.

Table 1. The parameters of the quadrotor DJI.

Parameter	Definition	Value
m	Mass	1.5 kg
l	The distance from the center of mass to a motor	0.225 m
g	Gravitational acceleration	$9.81\ \text{m/s}^2$
$I_{xx} = I_{yy}$	Moments of inertia around x, y axis in b-frame	$1.43 \times 10^{-5}\text{kg} \cdot \text{m}^2$
I_{zz}	Moments of inertia around z axis in b -frame	$2.89 \times 10^{-5}\text{kg} \cdot \text{m}^2$

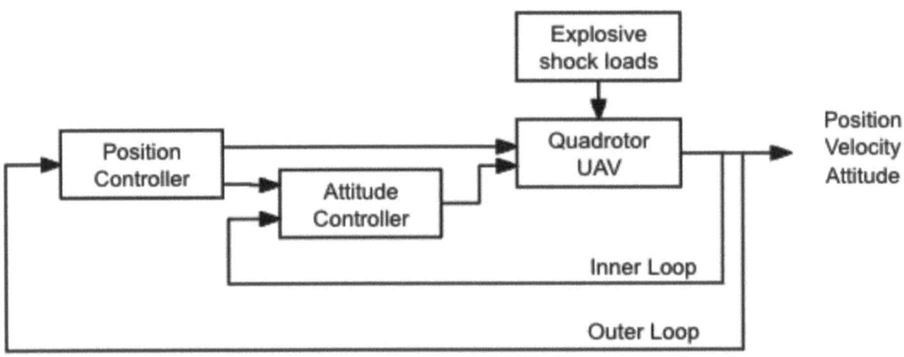

Fig. 5. The block diagram of position/attitude control structure for Quadrotor UAV with explosive shock loads.

3 Simulation Results and Discussion

The simulation results presented in Figs. 6, 7, 8, 9, 10, 11, 12 and 13 capture the comprehensive dynamic response of a quadrotor UAV when subjected to explosive shock loads during stationary hovering. Prior to the application of explosive forces, the aircraft executed a precise maneuver to establish a stabilized hover at position [10, 10, 10], maintaining this condition for 15 s to ensure consistent initial conditions. At this precisely controlled moment (t = 15 s), the system was subjected to explosive shock loads from specialized mechanical devices, with both horizontal and oblique loading configurations systematically examined to evaluate directional effects on aircraft dynamics.

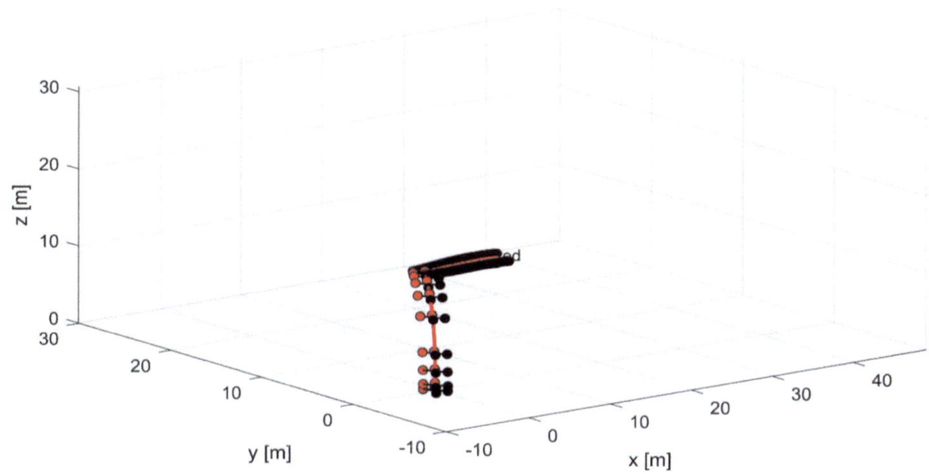

Fig. 6. The dynamic responses of trajectories to horizontal explosive shock loads.

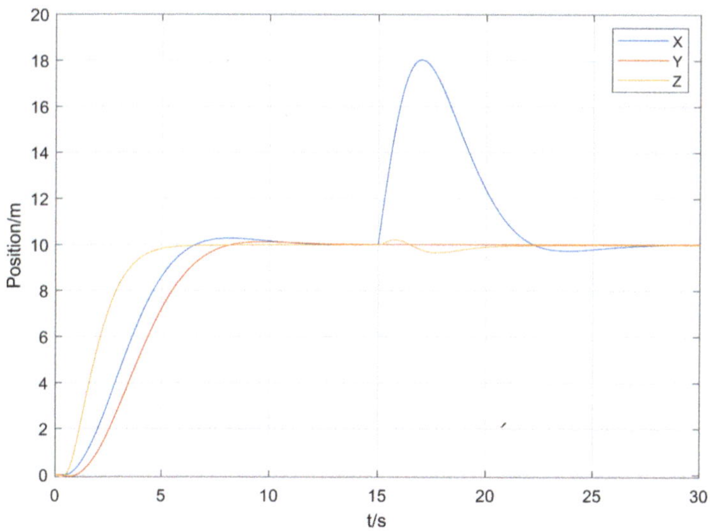

Fig. 7. The dynamic responses of positions to horizontal explosive shock loads.

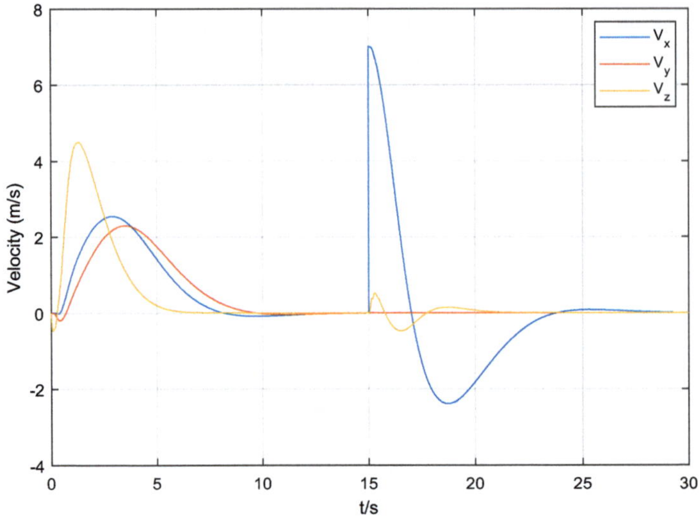

Fig. 8. The dynamic responses of velocities to horizontal explosive shock loads.

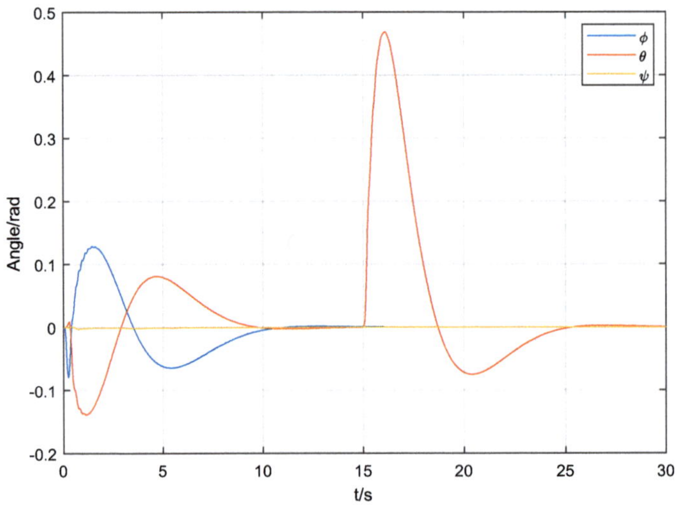

Fig. 9. The dynamic responses of Euler angles to horizontal explosive shock loads.

The simulation results presented in Figs. 6, 7, 8 and 9 reveal significant dynamic responses of the quadrotor UAV when subjected to horizontal explosive shock loads from the specialized mechanical device. The analysis demonstrates pronounced perturbations in all measured parameters, with particularly dominant effects observed in the x-direction translational dynamics. Displacement and velocity components along the x-axis exhibit the most substantial deviations, while attitude responses show a preferential sensitivity in roll angle compared to pitch and yaw angles.

This directional response pattern directly results from the force application geometry, where the horizontally oriented explosive shock load creates a dominant excitation along the x-axis. The observed roll moment likely stems from asymmetries in the blast wave interaction with the quadrotor's airframe and rotor configuration. Through implementation of the modified control architecture (Fig. 5), the system demonstrates effective disturbance rejection capabilities, achieving complete stabilization within a consistent 10 s recovery period. This recovery performance indicates that while the explosive shock induces substantial transient perturbations, the control system maintains sufficient authority to restore both positional and attitudinal equilibrium within an operationally relevant timeframe.

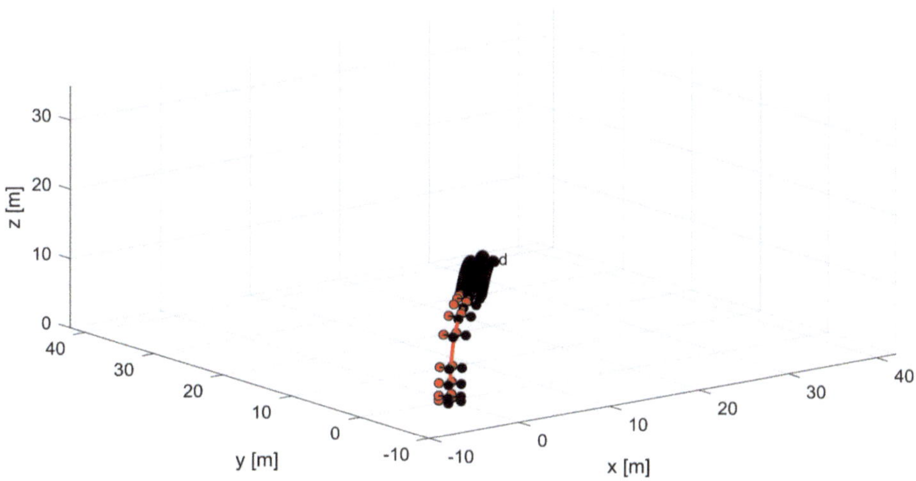

Fig. 10. The dynamic responses of trajectories to oblique explosive shock loads.

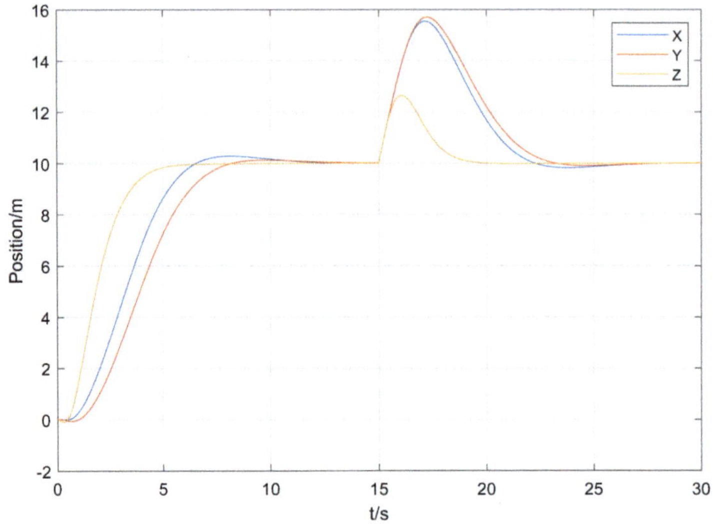

Fig. 11. The dynamic responses of positions to oblique explosive shock loads.

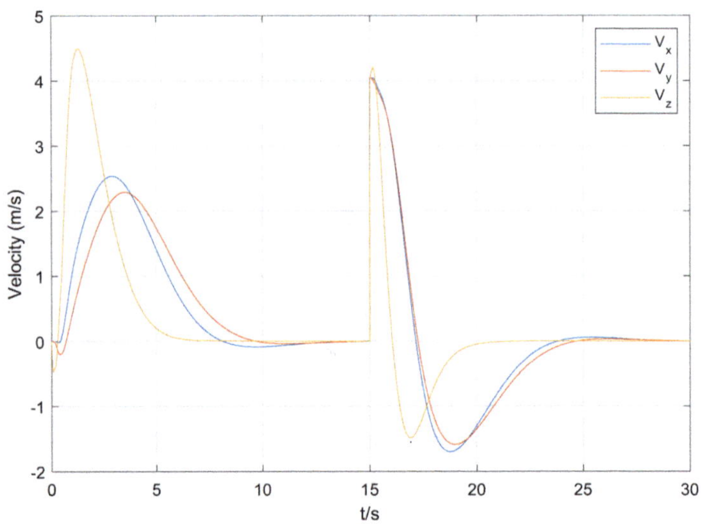

Fig. 12. The dynamic responses of velosities to oblique explosive shock loads.

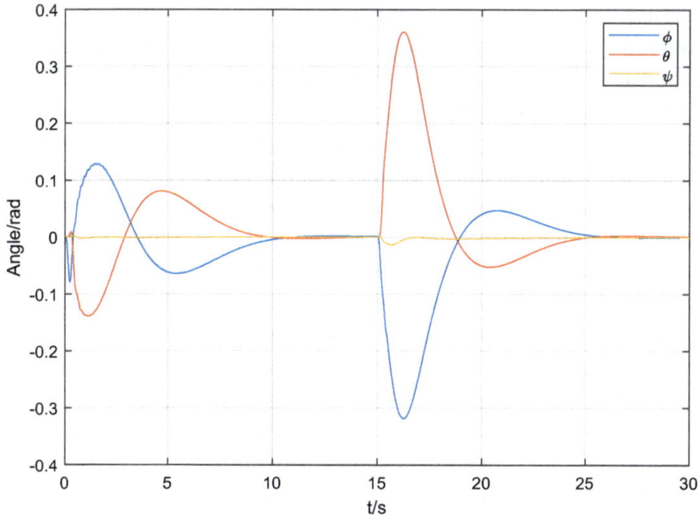

Fig. 13. The dynamic responses of Euler angles to oblique explosive shock loads.

The simulation results presented in Figs. 10, 11, 12 and 13 demonstrate that oblique explosive shock loads from the specialized mechanical device induce significantly attenuated dynamic responses in the quadrotor UAV compared to horizontal loading conditions. When subjected to oblique impacts, the UAV experiences distributed perturbations across all three spatial axes (x, y, and z), with each component exhibiting reduced intensity relative to the dominant x-direction response observed during horizontal loading. This attenuation effect stems from the vector decomposition of the explosive force, which effectively distributes a single concentrated impact into three orthogonal components of lower magnitude.

The same force distribution principle governs the UAV's angular response, with Euler angles showing proportionally reduced deviations under oblique loading. Notably, despite these mitigated perturbations in both translational and rotational degrees of freedom, the system maintains a consistent stabilization time of approximately 10 s. This temporal consistency reveals the control architecture's capability to simultaneously manage multi-axis disturbances, demonstrating its robustness in restoring flight stability regardless of impact directionality. The observed response characteristics suggest that while impact orientation affects perturbation magnitudes, the control system's stabilization performance remains largely invariant to load direction when properly tuned.

4 Conclusion

This study presents a systematic investigation into the dynamic response characteristics of quadrotor UAVs when subjected to explosive shock loads generated by specialized mechanical devices. Through comprehensive numerical modeling and simulation analysis, the research examines both horizontal and oblique blast loading scenarios, with particular focus on their respective effects on UAV displacement characteristics, velocity profiles, and attitude dynamics. The simulation results, presented in Figs. 6 through 13, reveal fundamentally different response patterns under these distinct loading conditions.

When exposed to horizontal explosive shock loads, the UAV exhibits significant dynamic perturbations across all measured parameters, with the most pronounced effects observable in x-direction displacement. The system's response demonstrates that through proper control system adjustments, the UAV can effectively recover stable flight conditions within a consistent 10 s timeframe. In contrast, oblique shock loading scenarios produce markedly attenuated dynamic responses due to the spatial distribution of impact energy across all three principal axes. While this energy distribution results in reduced peak displacement amplitudes and angular deviations, the required stabilization period remains essentially unchanged at approximately 10 s, confirming the robustness of the current control architecture in managing both translational and rotational dynamics under transient disturbance conditions.

These findings provide valuable insights into the directional dependence of blast impact effects on UAV flight dynamics and the effectiveness of existing control strategies in mitigating such disturbances. The study makes significant contributions to the field by developing a validated simulation framework for blast impact analysis, quantitatively characterizing directional effects on flight stability, and demonstrating control system performance under extreme loading conditions. The results have direct implications for enhancing UAV operational capabilities in explosive environments, particularly for military reconnaissance, explosive ordnance disposal, and emergency response applications.

Future research directions should prioritize experimental validation to verify simulation results and refine model accuracy, as well as the development of advanced adaptive control architectures to further optimize stabilization performance. Such developments would enable more effective integration between mechanical design improvements and control system capabilities, ultimately leading to more robust UAV systems capable of operating reliably in high-threat environments involving explosive shock loads. The current work establishes a solid foundation for these subsequent investigations while providing practical insights for engineers and researchers working on UAV blast resilience.

Acknowledgment. This work was supported by the State Scholarship Fund (No. 202106840005) of the China Scholarship Council.

References

1. Clark, R.M.: Evolution of Uninhabited Combat Aerial Vehicles (UCAV). Air University Press (2000)
2. Cai, Z., Selezneva, M.S., Yang, M.: Dijkstra algorithm based minimum acceleration/snap quadrotor UAV trajectory planning. J. Phys. Conf. Ser. **2746**(1), 012028 (2024)
3. Ding, Y., Zhou, K., He, L., Yang, H.: Experimental research on the muzzle response characteristics of small unmanned ground vehicles with small arms. Proc. Inst. Mech. Eng. Part C J. Mech. Eng. Sci. **236**(9), 4660–4670 (2022)

4. Yang, M., Lin, J., Li, J., Cai, Z., Ren, H., Liu, Q.: Research on the measurement device of UAV firearm unlocking force. In: Proceedings of the Fourth International Conference on Mechanical Engineering, Intelligent Manufacturing, and Automation Technology (MEMAT 2023), pp. 967–973. SPIE (2024)
5. Yang, M., Cai, Z., Lin, J., Li, J., Ren, H., Liu, Q.: Adaptability analysis on the measurement device of UACV special machinery unlocking force. In: Proceedings of the International Conference on Mechatronic Engineering and Artificial Intelligence (MEAI 2023), pp. 202–207. SPIE (2024)
6. Prasad, P.S., Asokan, S., Nayak, J.: Measurement and mitigation of recoil forces of rifles on an unmanned aerial vehicle. ISSS J. Micro Smart Syst. **12**(1), 9–18 (2023)
7. Yongjie, Z., et al.: A review of impact tests of light and small UAV. J. Phys.: Conf. Ser. **1786**(1), 012039 (2021)
8. Chen, C., et al.: Research on dropping test and numerical simulation for unmanned aerial vehicle. Int. J. Crashworthiness **27**(4), 1205–1221 (2022)
9. Carlucci, D.E.: Ballistics: Theory and Design of Guns and Ammunition. CRC Press, Boca Raton (2007)
10. Quan, Q.: Introduction to Multicopter Design and Control. Springer (2017)
11. Cömert, C., Kasnakoğlu, C.: Comparing and developing PID and sliding mode controllers for quadrotor. IJMERR 194–199 (2017)
12. Lee, K.U., Choi, Y.H., Park, J.B.: Position control of a quadrotor using state transformation technique. IJMERR 489–494 (2018)

Analysis of Two Oscillating Airfoil Encountering Sinusoidal Motion

Ruofan Wang(✉)

Peking University, Beijing, China
2201213367@stu.pku.edu.cn

Abstract. When two airfoils are moving in a complex way, the wake of the before airfoil has a great influence on the movement of the after airfoil. The scenario studied in this paper is that two NACA 2412 airfoils are oscillating while making a sinusoidal motion, they have the same sinusoidal motion frequency and oscillation frequency. ANSYS CFX was used for numerical simulations to study the effect of the before airfoil on the after airfoil. The distance between the two airfoils affect the two airfoils' movement are discussed and compared. Finally, some results of the setting conditions are showed. The pressure contours, velocity vector illustrations and picture of lift coefficient and drag coefficient show some valuable information.

Keywords: Two airfoils · Complex motion · Dynamic mesh

1 Introduction

The study of unsteady aerodynamics of airfoils in the case of incompressible flow, particularly at moderate Reynolds number, has become increasingly important, especially in the context of micro-air vehicles (MAVs) and unmanned air vehicles (UAVs) [1]. Fixed-wing MAVs frequently navigate through restricted environments and must perform intricate maneuvers during flight. The airfoil oscillates during the set sinusoidal motion, and the wake of the before airfoil will have a great influence on the after airfoil, so it is necessary to investigate the interactions of the before airfoil and the after airfoil.

Munk developed the thin airfoil theory [2], Brinbaum [3] and Glauert [4] which is to estimate the lift coefficient of a certain angle of attack (AOA) airfoil. Wagner [5] provided the solution of an airfoil experiencing a sudden change in AOA. Different geometric configurations between the airfoils demonstrate a notable enhancement in the lift-to-drag ratio of the downstream airfoil, reaching up to 77% when compared to single airfoil arrangements. N.Hosseini et al. delved into optimization methods to determine the optimal configuration of two airfoils placed in tandem within an incoming flow at a low Reynolds number of 30000, based on chord length. The upstream airfoil underwent pitching at a fixed frequency, while the downstream airfoil maintained a constant AOA [6]. Thierry M.Faure et al. investigated the characteristics of a two-airfoil arrangement by exploring variations in AOA and distances between the airfoils. They conducted their study by adapting a discrete-time vortex numerical method, taking into accountant the

T. Zielinska (Ed.): PEAE 2025, LNNS 1679, 19–30, 2026.
https://doi.org/10.1007/978-3-032-08221-3_2

potential variations that arise from a given airfoil and Reynolds number [7]. Zikang Zhou et al. placed a semicircular object in front of the NACA 0012 airfoil to study the interaction of the airfoil with the surrounding air under the action of the Kármán vortex street, which can applied to the design of robotic fish [17]. Ahmed Osama Mahgoub et al. proposed a numerical framework which can help investigate the interaction between the isotropic turbulence and airfoils, their research focused on the interaction of airfoil's leading edge and the turbulence. The result is that the streamwise and wall-normal components are the most affected velocity components [18]. Jiaxin Hou et al. proposed a model to study the nitrogen condensation over airfoils, they used different airfoils to test the plausibility of this model. In the end, they verified the effectiveness of this model in this certain environment [19]. Saransh Abbey et al. numerically investigated the aerodynamic characteristics of the tandem airfoil arrangements of the Wortmann FX 63–137 [20]. Thierry et al. proposed an adaptation of a discrete-time vortex numerical method to study the features of a two-airfoil arrangement with variations in the angle of attack and distances between the airfoils [21]. The situation of two airfoils moving at the same time is becoming more and more common in practical applications, so it is necessary to do some simulation studies on this situation.

Nowadays, with the maturity of artificial intelligence and neural network technology, more and more scholars have begun to use these methods to achieve fluid simulation, The airfoil around the airfoil was modeled and optimized to maximize the lift to drag ratio by Yubiao Sun et al. by using a physical-informed neural network (PINN) the airfoil [14]. Chanwook Park et al. proposed a network based on interpolation theories and tensor composition, the interpolating neural network (INN), which solve the problem of too many training parameters in the past neural network [15]. Kuijun Zuo et al. proposed a deep neural network framework, using the NACA series of airfoils for experiments, and finally made the simulation faster [16].

In this problem, the shape of the computational domain of the fluid changes over time due to the deformation of the boundary of the solid structure, so it is necessary to update the mesh in the computational domain after the boundary moves or is deformed. In order to simulate the unsteady flow field with moving or deformed boundary accurately and effectively, the dynamic grid technique is an essential key technique in numerical simulation. The mesh deformation method updates the mesh in the computing domain according to the boundary movement or deformation through mesh deformation, mesh reconstruction and other methods, so as to adapt the change of the shape of the computing domain by the boundary movement or deformation. In the problems of the relative motion or deformation of multiple bodies and the irregular deformation of the variant aircraft, it is very difficult to generate a complex shape and uniform fitted structure mesh, and even if it is barely generated, the quality of the generated mesh cannot be guaranteed. Therefore, many scholars have developed multi-block docking structure meshes [8, 9], multi-block lap structure meshes [10], and overlapping mesh generation techniques, and developed corresponding mesh deformation methods for different mesh types, among which the most commonly used in fluid-structure interaction simulation are overlapping structure mesh generation technology and mesh deformation method [11–13]. By dividing the calculation region into multiple subregions, the overlapping mesh generation technology can generate structural meshed on each subregion. The subregion meshes can be overlapped in whole or in part, and information can be exchanged through the overlapping parts without strict docking. Therefore, the difficulty of mesh generation is greatly simplified. In addition, the grid in each subregion can move and deform independently, so it is easy to realize the update of the computational mesh. Therefore, the overlapping mesh is widely used to solve the unsteady flow field simulation problems in complex regions, such as the delivery of aircraft external objects and the deformation of airfoils.

The rest of the paper is organized as follows: In Sect. 2, we give the theoretical basis for numerical simulation of this model. In Sect. 3, we present the computational domain of the model, the meshing results of the model, and the computational simulation settings in ANSYS CFX in Sect. 4, we list the results of the initial simulation and compare the results by changing the distance between the two airfoils. In Sect. 5, we summarize and discuss all the work and suggested the directions for further research.

2 Theoretical Description

For inviscid flows, ANSYS CFX solves the Euler equations, which contains mass conservation equation, momentum conservation equation and energy conservation equation. Among the threes equations, the mass conservation equation has the same form for laminar flows, the other two equations are reduced because of the molecular diffusion.

Mass Conservation Equation

$$\frac{\partial \rho}{\partial t} + \nabla \cdot \left(\rho \vec{v}\right) = S_m \tag{1}$$

The source S_m is mass added to the continuous phase from the dispersed second and any user-defined source.

Momentum Conservation Equations

$$\frac{\partial}{\partial t}(\rho \vec{v}) + \nabla \cdot (\rho \vec{v} \vec{v}) = -\nabla p + \rho \vec{g} + \vec{F} \tag{2}$$

where p is the static pressure and $\rho \vec{g}$ and $\vec{F}$ are the gravitational body force and external body forces respectively. $\vec{F}$ also contains other model-dependent source terms such as porous-media and user-defined sources.

Energy Conservation Equation

$$\frac{\partial}{\partial t}(\rho E) + \nabla \cdot (\vec{v} \cdot (\rho E + p)) = -\nabla \cdot \left(\sum\nolimits_j h_j J_j\right) + S_h \tag{3}$$

where E represents the total energy, $h_j J_j$ is the product of heat flux and mass, S_h contains other model-dependent source.

In the solving process of this paper, we employed the shear-stress transport (SST) $k - \omega$ model, which was developed based on two existing models, standard $k - \varepsilon$ model and standard $k - \omega$ model. Below are descriptions of these models.

Shear-Stress Transport (SST) $k - \omega$ Model
The SST $k - \omega$ model has a similar form to the standard $k - \omega$ model. The turbulence kinetic energy, *k*, and the specific dissipation rate, ω, are obtained from the following transport equations:

$$\frac{\partial}{\partial \mathrm{t}}(\rho \mathrm{k}) + \frac{\partial}{\partial \mathrm{x_i}}(\rho \mathrm{k u_i}) = \frac{\partial}{\partial \mathrm{x_j}}\left(\Gamma_{\mathrm{k}}\frac{\partial \mathrm{k}}{\partial \mathrm{x_j}}\right) + \tilde{\mathrm{G}}_{\mathrm{k}} - \mathrm{Y_k} + \mathrm{S_k} \tag{4}$$

and

$$\frac{\partial}{\partial t}(\rho\omega) + \frac{\partial}{\partial x_i}(\rho\omega u_i) = \frac{\partial}{\partial x_j}\left(\Gamma_{\omega}\frac{\partial \omega}{\partial x_j}\right) + A \tag{5}$$

where $A = G_\omega - Y_\omega + D_\omega + S_\omega$, $\tilde{\mathrm{G}}_{\mathrm{k}}$ represents the generation of turbulence kinetic energy due to mean velocity gradients. G_ω represents the generation of Γ_{k} and Γ_ω represent the effective diffusivity of k and ω, respectively. $\mathrm{Y_k}$ and Y_ω represent the dissipation of k and ω due turbulence. D_ω represents the cross-diffusion term, $\mathrm{S_k}$ and S_ω are user-defined source term.

3 Numerical Simulation

The computational domain set for two airfoils is shown in Fig. 1, both airfoils are NACA 2412 in shape. The entire analysis domain is 12c in length, 6c in width, and 0.1c in thickness, where c represents the chord length of the airfoil.

The distance between the left edge of the first airfoil and the left boundary of the domain is 2c. The distance between two airfoils is 3c. We have confirmed that these dimensions for the computational domain are large enough not to have a measurable effect on the aerodynamic coefficients. The left boundary is defined as a pressure inlet, the right boundary is defined as pressure outlet. The top and bottom boundaries are defined as opening. The front and behind boundaries are defined as interface. The surfaces of the two airfoils are defined as slip wall.

We use Ansys ICEM to generate the initial computational mesh, whose geometrical details are shown in Fig. 2. In the process of meshing, the entire computing domain is first divided into 15 blocks, in which the long side is divided into 5 segments and the short side is divided into three segments. In the region near the two airfoils is divided into 60 blocks in the x direction and 40 blocks in the y direction. The area of the middle part of the two airfoils is divided into 80 blocks in the x direction. Both the upper left and lower left areas are divided into 40 blocks in the x and y directions. In the z direction, the computational domain is divided into 8 blocks. Before the mesh is smoothed, a small area around the airfoil is a rectangle, and its four vertices are connected to the four selected points on the airfoil. They are the two points of the leading edge and the two points of the trailing edge. For the trailing edge, if directly select the points of the original shape,

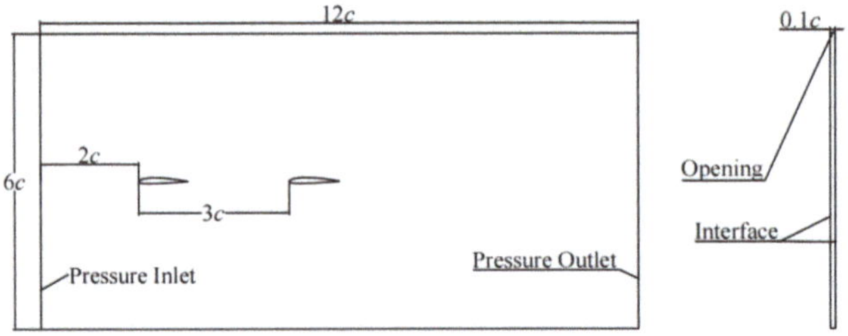

Fig. 1. Computational domain for two NACA 2412 airfoils

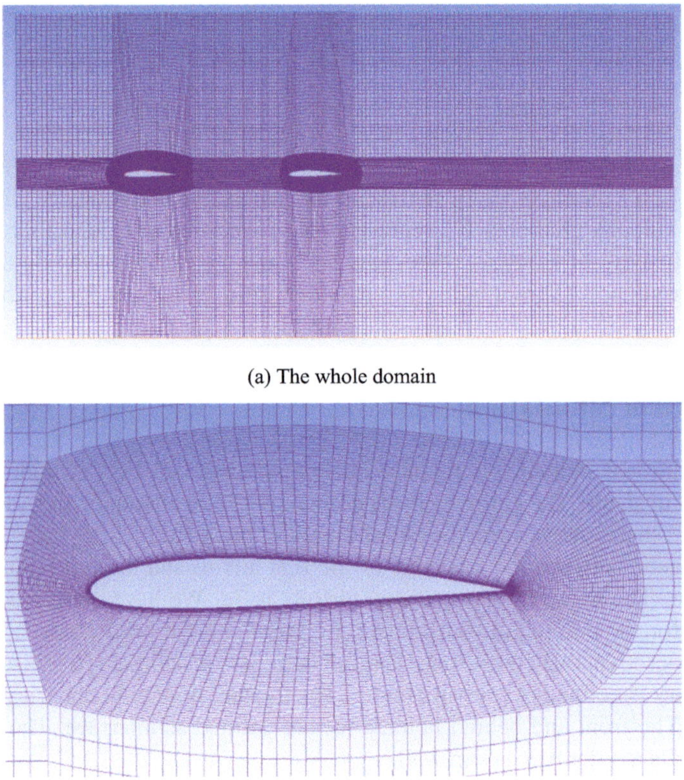

(a) The whole domain

(b) The after airfoil domain

Fig. 2. Computational mesh for the two airfoils

the quality of mesh will be very bad. So we removed some points on the trailing edge so that the airfoil shape at the trailing edge is a straight line in a small range. Both the upper right and lower right are divided into 160 blocks in the x direction. The final average mesh quality is 0.951. Both the upper right and lower right are divided into 160 blocks in the x direction. This results in more sparse meshes in unimportant areas, reducing unnecessary computational efforts. The transition areas from the airfoil to the regular hexahedral meshes near the airfoil are divided into denser and denser as they get closer to the airfoil, which makes the calculation results more accurate.

An ANSYS CFX solver, reliant on pressure calculations, is employed to pursue a dynamic solution. Air serves as the fluid medium, with density and dynamic viscosity configured to their default settings. The reference pressure is established at 0 atm, with stationary domain motion and specified regions of motion for mesh deformation. Total energy is selected for heat transfer, while the turbulence model opted for is shear stress transport.

For the inlet boundary we sat the relative pressure as 101.325 kPa and the static temperature is 298 K. We sat the relative pressure of outlet boundary as 90 kPa. The pressure values of the upper and lower boundaries are set to 90 kPa, and the static temperature is 288 K, the boundary type of these two surfaces is Opening. The boundary type of the front and back surfaces is Interface. The set of Interface is periodic, and the model is translational periodicity. We set the residual calculation method to RMS and stop the calculation when RMS is less than 1e-4. For the mesh deformation, we set the mesh deformation option to Regions of Mesh Specified, and the displacement relate to the previous mesh. The material of the fluid region is set to Air Ideal Gas. The relative pressure of top and bottom boundaries are also sat as 90 kPa. The computational domain sat up in ANSYS CFX is shown in Fig. 3.

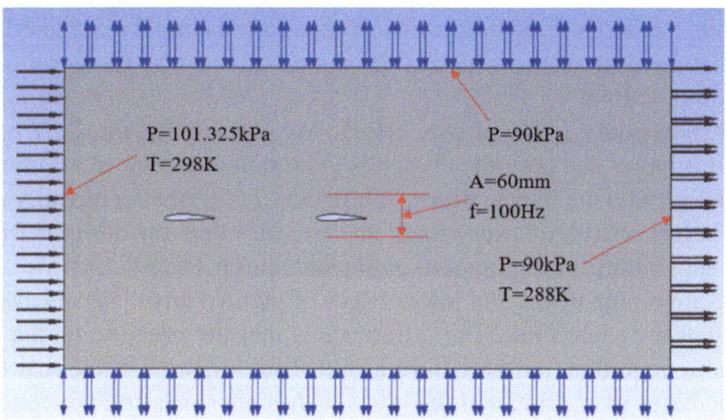

Fig. 3. The analysis domain for two NACA 2412 airfoils

The amplitude of the sinusoidal movement of the two airfoils is sat to 60 mm, the frequency is set to 100 Hz, and the oscillating movement frequency is also 100 Hz. The expression for the sinusoidal motion of the two airfoils is shown below.

$$\begin{aligned} y_1 &= sin(2\pi ft) \\ y_2 &= sin\left(2\pi ft + \tfrac{\pi}{2}\right) \end{aligned} \quad (6)$$

4 Results and Discussion

Output the above set file and run it with CFX-Solver Manager to obtain the results of before airfoil lift coefficient and drag coefficient together with the results of after airfoil drag and lift are shown in Fig. 4. In all of the following figures.

$$T^* = tf \tag{7}$$

where $f = 100\,Hz$ so T^* is a dimensionless.

$$C_L = \frac{F_L}{A \times q} \tag{8}$$

$$C_D = \frac{F_D}{A \times q} \tag{9}$$

where F_L, F_d are lift and drag respectively, A is surface area, here $A = 0.1\,m^2$, q is the dynamic pressure, here $q = P_i - P_o = 101.325 - 90\,kPa = 11.325\,\text{kPa}$.

From Fig. 4 we can see that the C_L and C_D of before airfoil show obvious periodicity. By contrast, influenced by the wave of before airfoil, the C_L and C_D of after airfoil are not periodic. At the same time, the phase of C_L movement in the after airfoil appears to lag behind the before airfoil. The results of the C_L performance are in line with our settings. While the drag off the before airfoil appears to be more disordered that the C_L, indicating that the drag of the after airfoil is more affected by the wake of the before airfoil under this setting.

Figure 5 shows the C_L/C_D of two airfoils. We can see that the C_L/C_D ratio of the before airfoil is larger and periodically tends to be stable, similar to a damping motion, this is also in line with our expectations. While the C_L/C_D of after airfoil, influenced by the wave of before airfoil, it is very small and we can't find any disciplinarian.

Figure 6 shows the pressure contour of this simulation. From T* = 1 we can conclude that the pressure on the upper and lower edges of the two airfoils is always decreasing or increasing at the same time. The difference is that the pressure on the after airfoil is always more huge than on the before airfoil. Besides, we can see that the pressure gradually increases over time, and the areas with higher pressure are concentrated on the upper and lower surfaces of the two airfoils.

From Fig. 7 we can see that the before airfoil has an increasing impact on the area behind it, especially in a wavy area behind the tail of the airfoil over time. As a result, the after airfoil head is getting faster and faster. And over time, we can see that the after airfoil tail speed is also increasing, gradually affecting the area behind the after airfoil more and more, just like the before airfoil. Based on this, we infer that when the two airfoils move at the same time, the after airfoil has a certain delay in velocity relative to the before airfoil, and its velocity characteristics will gradually converge with the before airfoil over time.

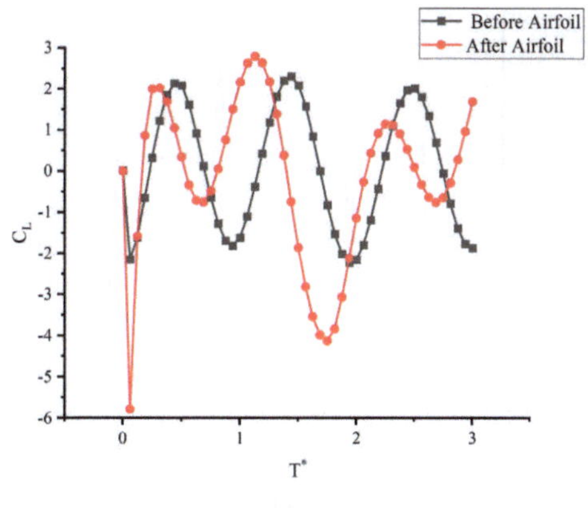

(a) C_L of two airfoils

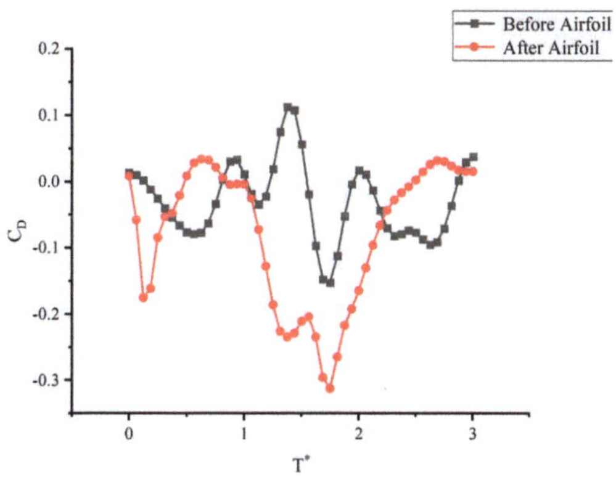

(b) C_D of two airfoils

Fig. 4. C_L and C_D results of before airfoil obtained by ANSYS CFD-Post and the distance between two airfoils is 3c and both of they are all in motion for three cycles

In the following, we will continue to discuss the effect of the distance between the two airfoils on the C_L and C_D of the after airfoil. Firstly, we set the distance between two airfoils as 4c, and set the length of whole domain as 13c. Since the calculation of motion and related physical quantities is more complicated by changing the distance between the two airfoils, it is easy to have a negative meshes during the calculation. So we've made some adjustments to the relevant parameters, the amplitude of the sinusoidal movement of the two airfoils is changed to 10mm, and the motion time of the simulation is set as a period of sinusoidal motion. Through this setting, the dynamic mesh in both cases can maintain high quality throughout the whole simulation process, and the results obtained are more reliable.

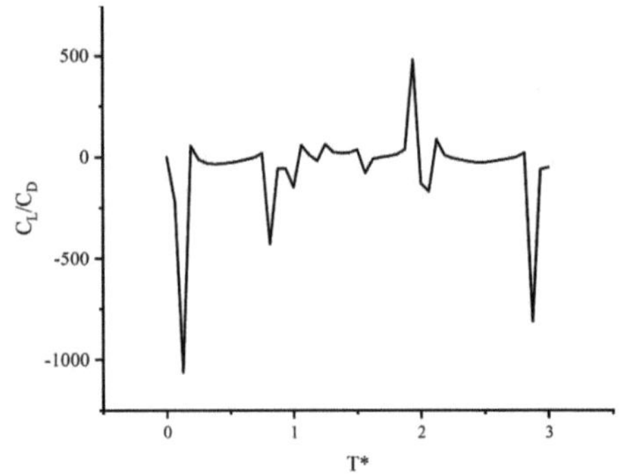

(a) C_L/C_D of before airfoil

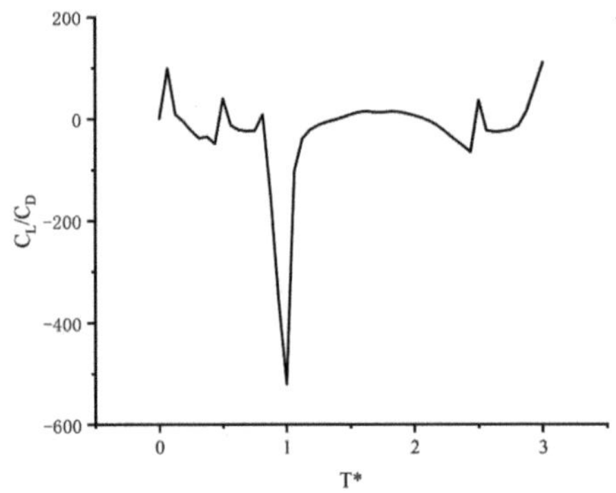

(b) C_L/C_D of after airfoil

Fig. 5. C_L/C_D results of before airfoil obtained by ANSYS CFD-Post and the distance between two airfoils is 3c and both of they are all in motion for three cycles

From Fig. 8 we can see that when the distance increased, the positive C_L and the negative C_D of after airfoil become greater. It proves that the impact of the before airfoil on the after airfoil is smaller, which is in line with our expectation.

From Fig. 9 we can see that the C_L/C_D of the before airfoil is still much larger that of the after airfoil when the distance is reduced. In addition, the C_L/C_D of the before airfoil also shows a periodic effect of gradual amplitude attenuation to a certain extent.

From Fig. 10, after the distance modified to 4c, compared to before the distance changed, the C_L/C_D of two airfoils are very close.

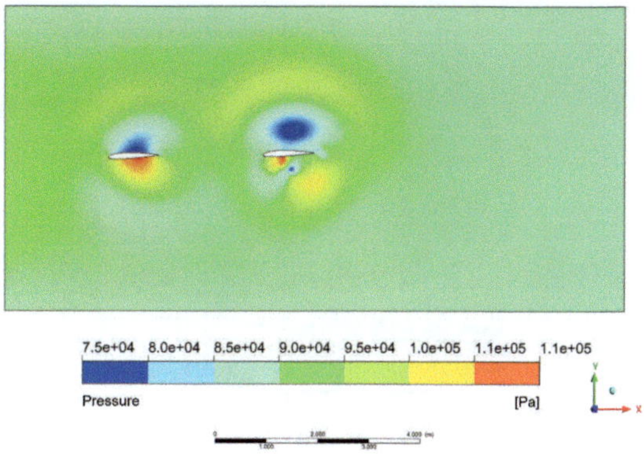

(a) $T^{*} = 0.5$

(b) $T^{*} = 1$

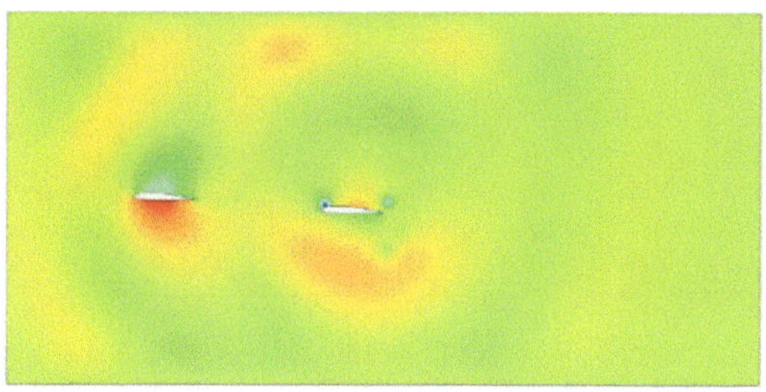

(c) $T^{*} = 1.5$

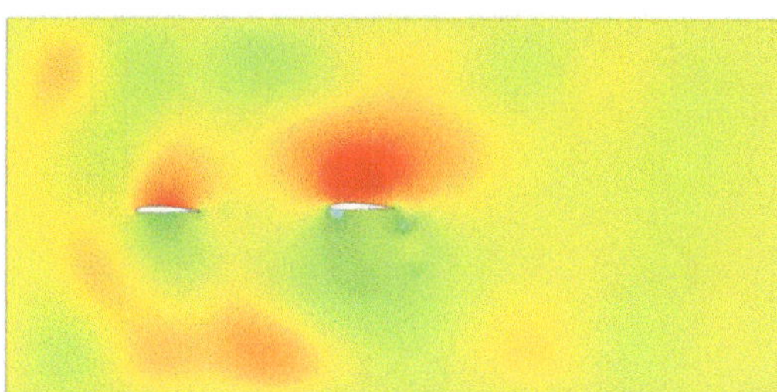

(d) $T^{*} = 2$

Fig. 6. Pressure results of the whole compute domain obtained by ANSYS CFD-Post and the distance between two airfoils is 3c and both of they are all in motion for three cycles

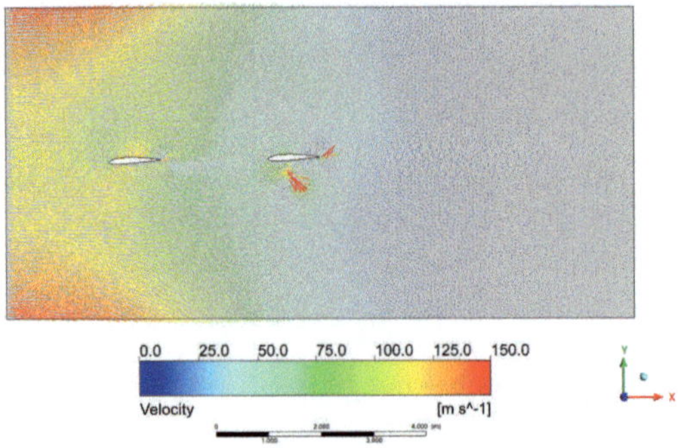

(a) $T^{*} = 0.5$

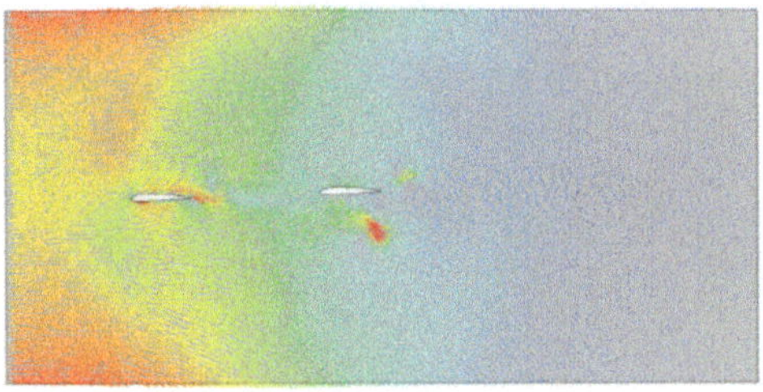

(b) $T^{*} = 1$

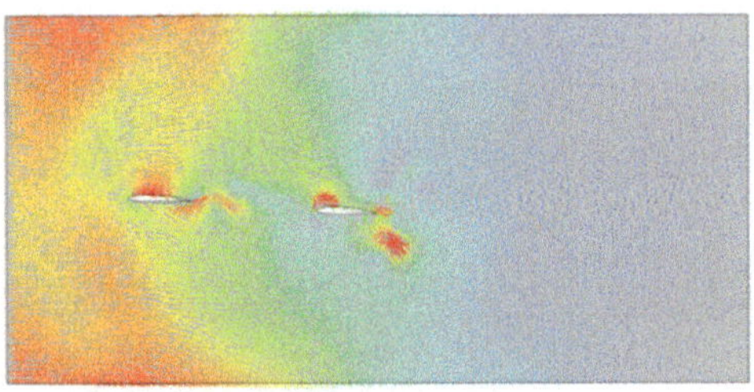

(c) $T^{*} = 1.5$

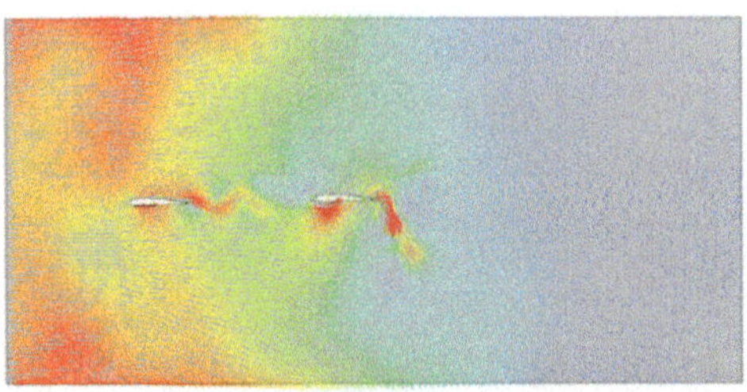

(d) $T^{*} = 2$

Fig. 7. Velocity results of the whole compute domain obtained by ANSYS CFD-Post and the distance between two airfoils is 3c and both of they are all in motion for three cycles

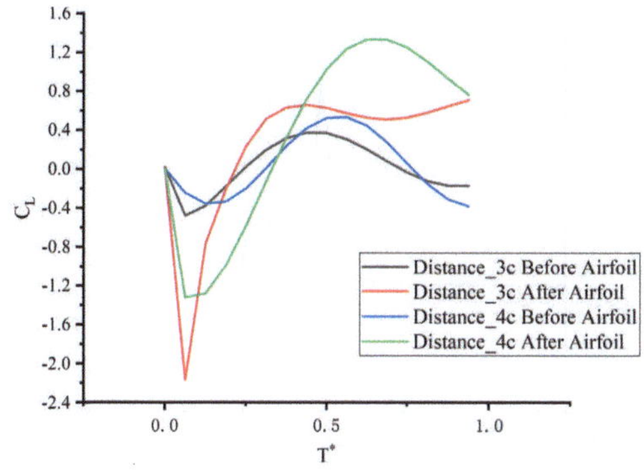

(a) Comparison of C_L

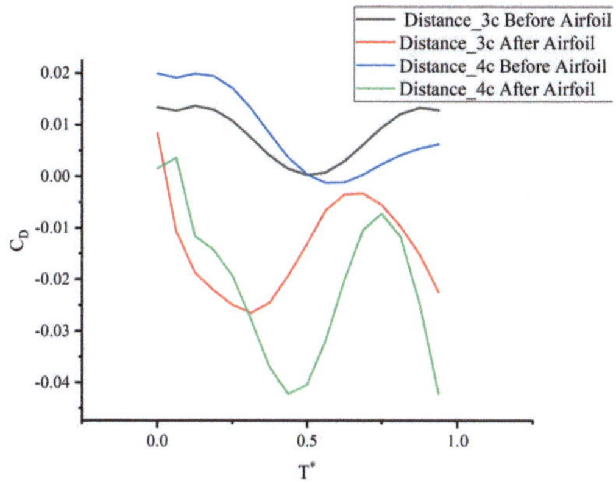

(b) Comparison of C_D

Fig. 8. Comparison of C_L and C_D results on the before airfoil when the distance between the two airfoils is 3c and 4c, respectively

When the distance between the two airfoils is 4c, after adjusting the amplitude of the sinusoidal motion to 10 mm the C_L C_D and C_L/C_D of the two airfoils are more reasonable. Therefore, in practical application, when the two airfoils move forward and backward, we should try to avoid large up and down movements, otherwise the movement will be very unstable.

5 Conclusion

In this paper, we use ANSYS CFX to simulate the complex motion of the before airfoil and after airfoil, both of which are in periodic sinusoidal and oscillating motion at the same time, with a certain phase difference between their sinusoidal motion. We analyzed the results of the C_L and C_D. To a more comprehensive analysis, we adjusted the distance between the two airfoils. It turned out that the distance between the two airfoils had little effect on the C_L/C_D when the range of changes was not large.

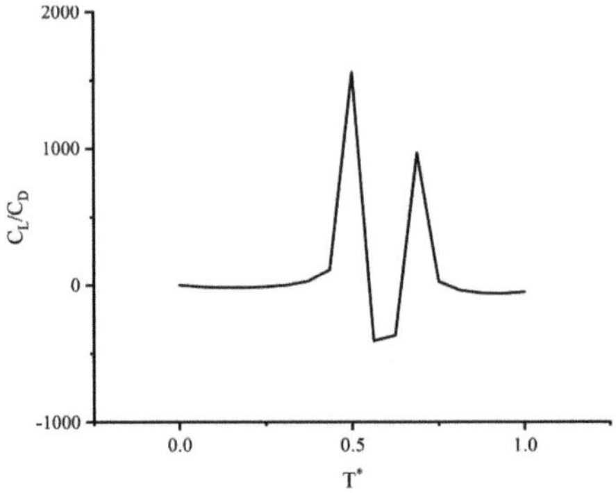

(a) C_L/C_D results of before airfoil when the distance is 4c

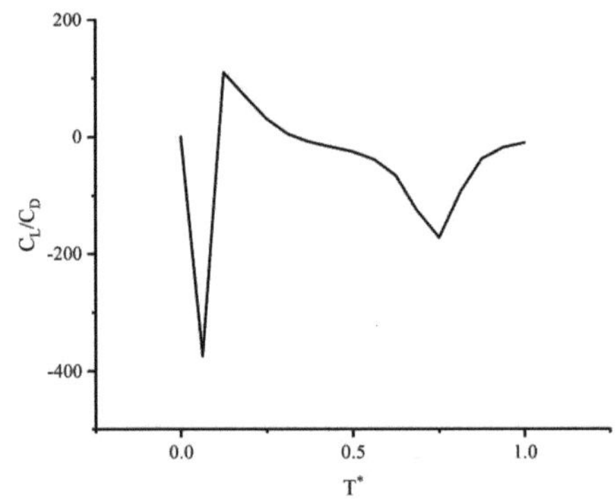

(b) C_L/C_D results of after airfoil when the distance is 4c

Fig. 9. C_L/C_D results on the before and after airfoil obtained by ANSYS CFD-Post and the distance between two airfoils is 4c and both of they are all in motion for one cycle

For the effect of distance, we would have liked to modify the distance between the two airfoils to 2c, 5c, 6c or something else. However, in the actual research process, we found that although the minimum mesh mass in the initial state has reached 0.835, a negative mesh soon appears in the calculation process, which makes the calculation process unable to continue. This may be due to weakness of ANSYS CFX's moving mesh algorithm for solving complex motions of two airfoils. In the following research, we want to develop a set of adaptive and powerful algorithm for the dynamic mesh algorithm for such problems. Motion related parameters such as the oscillation frequency of the two airfoils, the phase difference of the sinusoidal motion. Parameters related to the calculation domain, such as the pressure difference between the inlet and outlet, temperature, etc., are all aspects worth discussing.

In the follow-up study, in addition the distance between the two airfoils, we would like to continue to change some other parameters to study the interactions between the two airfoils.

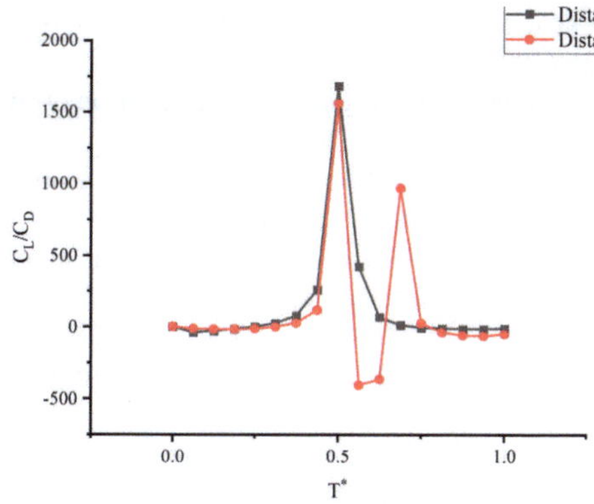

(a) C_L/C_D results of before airfoil when the distance is 3c, 4c respectively

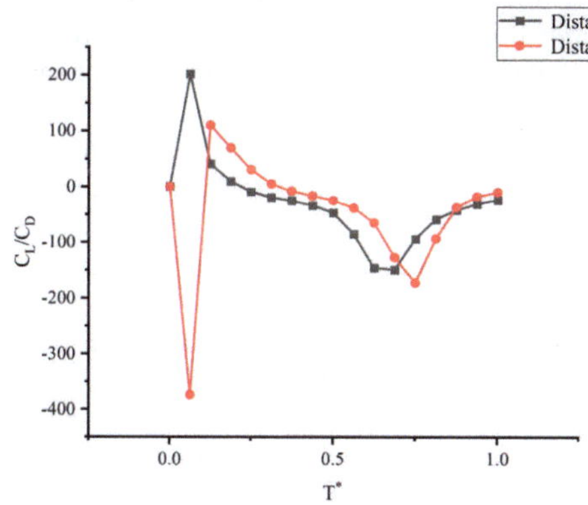

(b) C_L/C_D results of after airfoil when the distance is 3c, 4c respectively

Fig. 10. Comparison of C_L/C_D results on the before airfoil and after airfoil when the distance between the two airfoils is 3c and 4c, respectively

As mentioned in the introduction, for complex fluid simulation problems, methods such as artificial intelligence can effectively reduce the time required to solve them. Therefore, we will consider using deep learning methods [22, 23] to study this problem in more depth.

References

1. Mueller, T.J., DeLaurier, J.D.: Aerodynamics of small vehicles. Annu. Rev. Fluid Mech. **35**(1), 89–111 (2003)
2. Munk, M.: General theory of thin wing sections. Technical report, 142 NACA (1922)
3. Birnbaum, W.: Die tragende Wirbelfläche als Hilfsmittel zur Behandlung des ebenen Problems der Tragflügeltheorie. ZAMM J. Appl. Math. Mech. **3**(4), 290–297 (1923)
4. Glauert, H.: The Elements of Aerofoil and Airscrew Theory. Cambridge University Press, Cambridge (1926)

5. Wagner, H.: Über die Entstehung des dynamischen Auftriebes von Tragflügeln. Z. Angew. Math. Mech. **5**(1), 17–35 (1925)
6. Hosseini, N., Tadjfar, M., Abba, A.: Configuration optimization of two tandem airfoils at low Reynolds numbers. Appl. Math. Model. 828–846 (2022)
7. Liang, H., Wang, X., Zou, L., Zong, Z.: Numerical study of two-dimensional heaving airfoils in ground effect. J. Fluids Struct. 188–202 (2014)
8. Steinthorson, E., Liou, M.S., Povinelli, L.A.: Development of an explicit multiblock/multigrid flow solver for viscous flows in complex geometries. In: AIAA, pp. 93–2380 (1993)
9. Sheng, C., Taylor, L., Whitfield, D.: Multiblock multigrid solution of three dimensional incompressible turbulent flows about appended submarine configuration. In: AIAA, pp. 95–203 (1995)
10. Flores, J., Reznick, S.G., Holst, T.L., et al.: Transonic Navier-Stokes solution for a fighter-like configuration. In: AIAA, pp. 875–881 (1987)
11. Lijewski, L.E.: Comparison of transonic store separation trajectory predictions using the Pegasus/DXEAGLE and BEGGER Codes. In: AIAA, pp. 97–2202 (1997)
12. Wang Z J, Parthasarathy V. A fully automated Chimera methodology for multiple moving body problems. International Journal for Numerical Methods in Fluids. 919–938 (2000)
13. Yen, G.W., Baysal, O.: Dynamic-overlapped-grid simulation of aerodynamically determined relative motion. In: AIAA, pp. 93–3018 (1993)
14. Sun, U., Sengupta, U., Juniper, M.: Physics-informed deep learning for simultaneous surrogate modeling and PDE-constrained optimization of an airfoil geometry. Comput. Methods Appl. Mech. Eng. (2023). ISSN 0045–7825
15. Park, C., et al.: Engineering software 2.0 by interpolating neural networks: unifying training, solving, and calibration. arXiv:2404.10296v3 (2024)
16. Zuo, K., Ye, Z., Bu, S., Yuan, X., Zhang, W.: Fast simulation of airfoil flow field via deep neural network. Aerosp. Sci. Technol. 1270–9638 (2024)
17. Zhou, Z., Huang, S., Wang, Y.: Numerical simulation study of the fluid-structure coupling effects of Kármán vortex street on airfoil motion with varying thickness. Ocean Eng. 0029–8018 (2023)
18. Mahgoub, A.O., Jiang, C., Moreau, D., Doolan, C., de Silva, C.: A numerical framework to investigate isotropic turbulent inflow interacting with an airfoil's leading edge. Int. J. Heat Fluid Flow 0142–727 (2024)
19. Hou, J., Xiao, X., Xie, J., Chen, J.: Numerical investigation of nitrogen spontaneous condensation over airfoils in cryogenic wind tunnel. Case Stud. Thermal Eng. (2024). ISSN 2214–157X
20. Abbey, S., Dankhara, K., Senthilkumar, S.: Aerodynamic characteristics of flow over two unsymmetrical tandemly arranged airfoils – Numerical simulation. In: 3rd International Conference on Advances in Mechanical Engineering, p. 042033 (2020)
21. Aure, T.M., Dumas, L., Montagnier, O.: Numerical study of two-airfoil arrangements by a discrete vortex method. Theor. Comput. Fluid Dyn. **34**, 79–103 (2020)
22. Zhang, L., Lu, Y., Tang, S., Liu, W.K.: HiDeNN-TD: reduced-order hierarchical deep learning neural networks. Comput. Methods Appl. Mech. Eng. **389**, 114414 (2022)
23. Lu, Y., et al.: Convolution hierarchical deep-learning neural networks (c-hidenn): finite elements, isogeometric analysis, tensor decomposition, and beyond. Comput. Mech. 1–30 (2023)

Aerodynamic Design and Experimental Validation of Ailerons Downward Deflection in the Takeoff Configuration of a Civil Aircraft

Wenkui Wang, Kun Mao, Meihong Zhang(✉), and Huiwen Xu

Shanghai Aircraft Design and Research Institute, Jinke Road, Shanghai 200126, China
boaiwwk@sina.com, {maokun,zhangmeihong,xuhuiwen}@comac.cc

Abstract. This paper focused on a civil aircraft with the aim of improving the takeoff lift-to-drag ratio through the aerodynamic design of symmetric downward deflection of bilateral ailerons. The drag reduction benefits corresponding to various ailerons deflections were computed and validated via wind tunnel experiments. Additionally, the impacts of factors such as trim and gaps at the ends of the ailerons on the drag reduction benefits were analyzed. The relevant conclusions suggested that the influence of trim and gaps must be considered in the design of ailerons downward deflection. Based on these conclusions, the optimal deflection angle of ailerons was determined, and drag coefficient of the total aircraft was reduced by about 0.69%.

Keywords: ailerons downward deflection · lift-to-drag ratio · trim · gap · wind tunnel experiments

1 Introduction

During the takeoff and landing phases of an aircraft, due to the low speed, the high lift devices must be activated, and a large angle of attack must be maintained to achieve a higher lift coefficient. Unlike the landing configuration, which aims for high lift and high drag, the takeoff configuration seeks high lift with low drag, and the deflection angles of flaps and slats are smaller. Improving the takeoff lift-to-drag ratio can effectively shorten the takeoff running distance, and increase the climb gradient, thereby reducing fuel consumption during takeoff phase and enhancing commercial payload capacity. This improvement is of significant implications.

According to the classical lifting-line theory [1], the elliptical lift distribution can minimize the induced drag. Usually, the flaps of transport aircraft are located in the inner and middle areas of the wing. During the take-off phase, the deflection of the flaps causes a significant increase in the load within the inner wing area, and the lift distribution significantly deviates from the elliptical distribution. The ailerons are generally located on the outer area of the wing. For take-off configuration, symmetrically deflecting the ailerons on both sides of the aircraft at an appropriate angle can shift the wing load center outward, thus making the lift distribution closer to the elliptical distribution and effectively reducing the drag.

T. Zielinska (Ed.): PEAE 2025, LNNS 1679, pp. 31–42, 2026.
https://doi.org/10.1007/978-3-032-08221-3_3

Ailerons symmetrical downward deflection can be considered as a technology for variable camber wing. Variable camber technology is to control flaps, ailerons, spoilers or other control surfaces to deflect at different angles during various flight with the aim to reduce drag. In the 1990s, based on L-1011, NASA [2] conducted research on the drag reduction benefits of variable camber wing in civil aircrafts. Lyu and Martins [3, 4] utilized an adjoint optimization method to investigate the advantages of variable wing trailing edge camber for CRM. Guo et al. [5] employed the Free Form Deformation (FFD) method to design wing trailing edge camber variation for CRM. Niu et al. [6] combined the traditional aerodynamic optimization method based on Class-Shape Transformation (CST) with the variable camber technology of wing trailing edge, and obtained better optimization results compared to using the former alone. At present, variable camber technology has been applied on many types of civil aircraft. Boeing 787 and A350 [7, 8] use the variable camber technology of the wing trailing edge based on the Adaptive Dropped Hinge Flap (ADHF) to reduce drag during cruise phase. The ailerons of A320 and Boeing 777 aircraft can symmetrically downward deflect by 5° and 8° [9] respectively to improve the lift-to-drag ratio during takeoff phase.

Although there has been considerable research on variable camber wing technology, most of the studies focused on high-speed cruise phase of aircraft, with less attention paid to ailerons symmetric deflection during takeoff phase. In China, no wind tunnel experiments have been conducted to investigate the impact of ailerons downward deflection on improving the lift-to-drag ratio. This paper took a civil aircraft as the research object and conducted a ailerons deflection design for its takeoff configuration. The optimal deflection angle of ailerons was determined while considering the influence of the spanwise gap and the trim loss, then the wind tunnel experiment is conducted to validate it.

2 Numerical Methods and Validation

2.1 Numerical Methods

The computational resources required for solving Reynolds-averaged Navier-Stokes equations (RANS) are less, and the computational accuracy is acceptable, making it widely used in industrial applications. The integral form of the three-dimensional Navier-Stokes equation is:

$$\frac{\partial}{\partial t}\iiint_{\Omega} Q dV + \iint_{\partial\Omega} F(Q)\cdot \boldsymbol{n} dS = \iint_{\partial\Omega} G(Q)\cdot \boldsymbol{n} dS. \tag{1}$$

In (1), Ω is the control volume, $\partial\Omega$ represents the boundary of the control volume element, and Q represents the conserved variable,

$$Q = \begin{bmatrix} \rho & \rho u & \rho v & \rho w & \rho E \end{bmatrix}^T. \tag{2}$$

ρ is the density of fluids. u, v, w are the flow velocities in three directions. E is the total internal energy per unit mass of fluids. F(Q) is the inviscid flux term, and G(Q) is the viscous flux term.

In this paper, the open-source structured grid solver, CFL3D, was used to solve the three-dimensional RANS equations. Spalart-Allmaras turbulence model was adopted. Third-order upwind scheme was used to spatially discretize the flux. And the flux difference splitting (FDS) method was Roe scheme.

2.2 Validation Case

The three-element airfoil, 30P30N, was the case presented at the CFD Challenge Workshop held by NASA Langley in 1993. Its wind tunnel experiments were mainly completed in the low-turbulence wind tunnel at NASA Langley in the 1990s. To validate numerical methods' effectiveness in calculating the aerodynamic forces of three-dimensional high lift devices. 30P30N airfoil is stretched along spanwise and then used as a validation case. The free stream Mach number is 0.2, and the Reynolds number based on the chord length is 9×10^6.

Chord length of 30P30N is 1.0 m. The computational domain is a cuboid. And the boundary of computational domain is about 50c away from the wing. A C-H type multi-block structured grid is used. The first layer height of the boundary layer is about 3×10^{-6}c. The computational grid is shown in Fig. 1. The aerodynamic coefficients obtained from RANS are compared with the experimental values, as shown in Fig. 2. CL and Cm values of CFD are in good agreement with the experimental values in the linear region, but there is a noticeable deviation in the near stall region. Critical angle of attack predicted by CFD is larger than the experimental value. CD values computed are significantly larger, mainly because of the use of full turbulent computation. Although the computed values of CD deviate significantly, the difference between the computed and experimental values are stable, about 0.017, so they still have considerable reference value.

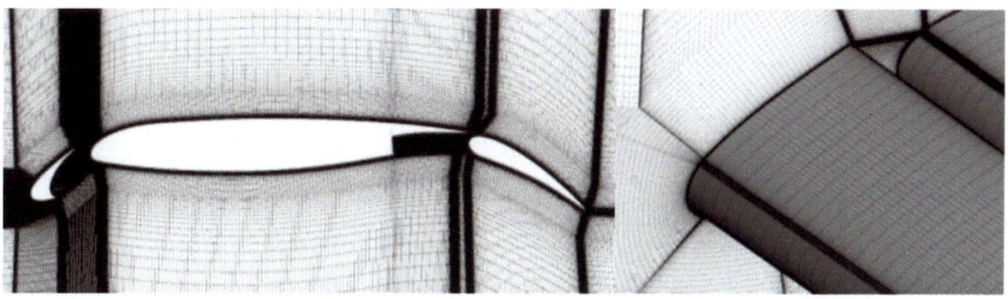

Fig. 1. Computation grid of 30P30N

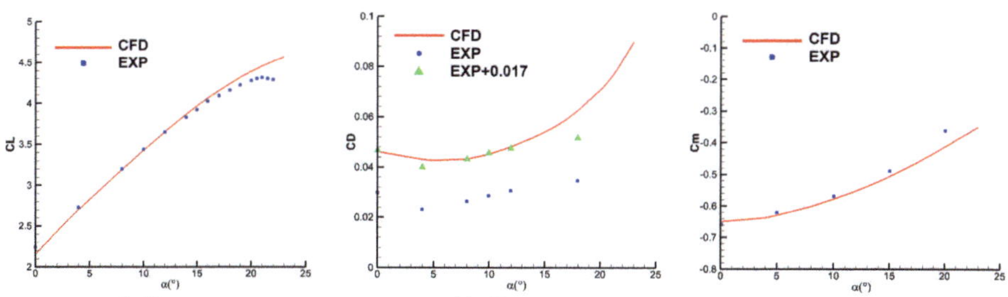

a) CL-α b) CD-α c) Cm-α

Fig. 2. Aerodynamic values of computation and experiments

In this paper, the ailerons deflection scheme of the civil aircraft mainly focuses on the linear region of the CL-α curve, and the stall characteristics predicted by CFD are relatively unimportant. In the linear region, $\alpha = 8°$ is taken as an example, aerodynamic coefficients (except CD, Table 1) and Cp distribution obtained from CFD are in good agreement with the experimental values (Fig. 3). CL, Cm and Cp distribution are computed accurately, and the deviation of CD computed values is stable. The computation methods have been validated and are applicable to ailerons deflection design.

Table 1. $\alpha = 8°$, computed and experimental values of aerodynamic coefficients

	CL	CD	Cm
CFD	3.196	0.04320	−0.601
EXP	3.201	0.02639	−0.593
Error	−0.16%	63.70%	1.35%

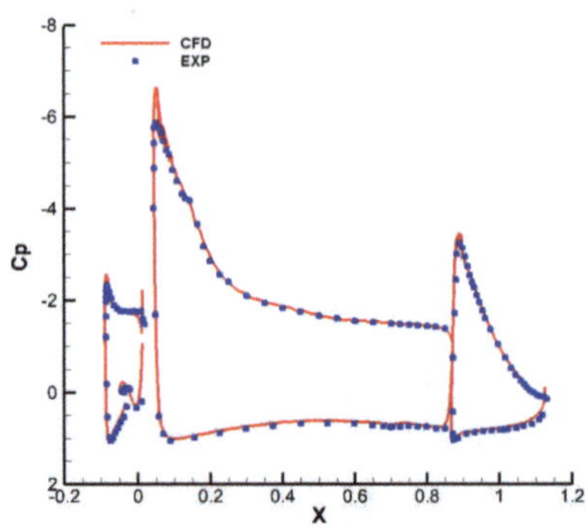

Fig. 3. $\alpha = 8°$, computed and experimental values of Cp distribution

3 Ailerons Deflection Design Based on CFD

3.1 Overview of the Model

The research object of this paper is a narrow-body civil aircraft, as shown in Fig. 4, which adopts a conventional twin-engine wing-mounted layout. The leading edge is equipped with slats as high lift devices, while the trailing edge features single-slot Fowler flaps. The outer wing section of the wing on both sides has an aileron. Usually, the ailerons on both sides deflect in different directions to obtain the lateral and directional control moment. During takeoff phase, the ailerons on both sides are symmetrically deflected downward at a certain angle, which can improve the lift-to-drag ratio. On this basis, the ailerons on both sides can be superimposed with different deflections, and still have lateral and directional control capabilities.

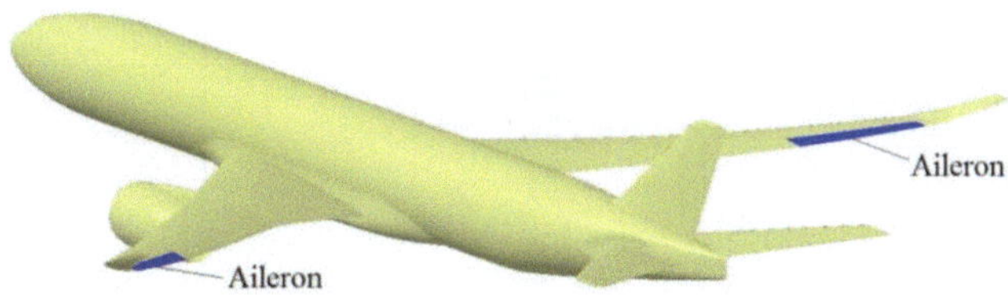

Fig. 4. Model of the civil aircraft

The design target of takeoff lift-to-drag ratio is mainly constrained by the climbing gradient in the second phase of takeoff. According to the relevant provisions of CCAR-25 on the take-off and landing performance, based on typical routes and typical payload, the typical lift coefficient CL_{target} of this civil aircraft in the second stage can be calculated. Using the ailerons deflection downward aims to improve the lift-to-drag ratio at $CL = CL_{target}$, while minimizing the loss of stall characteristics and control characteristics.

3.2 Calculation and Analysis of Different Ailerons Downward Deflection Angles

In the takeoff configuration, ailerons are deflected downward by different angles. For each ailerons deflection, multi-blocks structural grid is built for the whole aircraft (Fig. 5). When ailerons are deflected downward, transition surfaces are used to fill the gaps between the aileron and the fixed wing surface. The aerodynamic coefficients are calculated and analyzed for different ailerons deflection angles (Case 1–4) in the gapless configuration, as shown in Fig. 6. It is obvious that larger ailerons deflection angle leads to increased lift and nose-down moment.

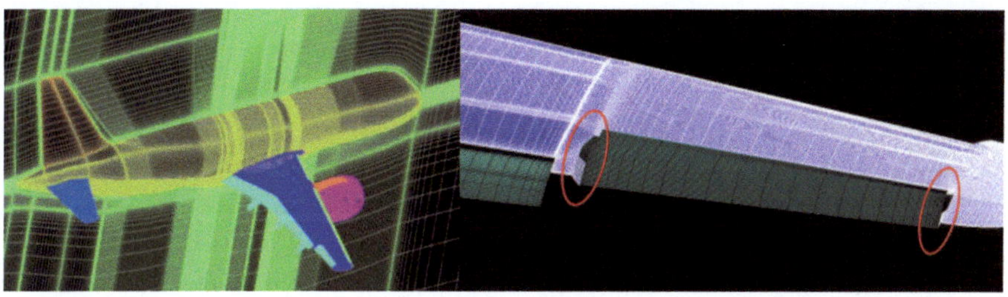

Fig. 5. Computation grid of ailerons downward deflection in takeoff configuration

Table 2 shows ΔCD corresponding to various ailerons angles. The configuration with $\delta_a = 0°$ is set as the baseline, and ΔCD of each deflection angle is shown as a percentage of the CD of the baseline.

Ailerons downward deflection causes a significant increasement in the nose-down moment. A larger nose-down moment results in a larger deflection angle of the horizontal tail, which causes additional drag that may offset drag reduction of ailerons deflection. Therefore, the effect of trimming on CD must be considered. Previous wind tunnel experiments have provided the relationship between horizontal tail deflection angle and

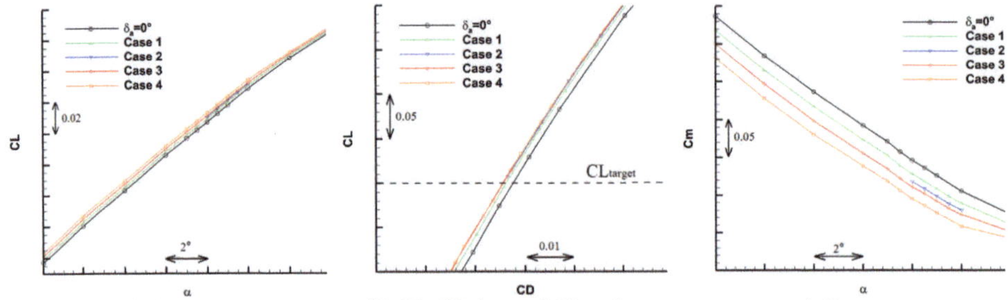

a) CL-α b) CL-CD (around CL_{target}) c) Cm-α

Fig. 6. Aerodynamic coefficients of takeoff configuration at different ailerons deflection angles

Table 2. CL = CL_{target}, ΔCD corresponding to different ailerons downward deflection angles

No.	δ_a	ΔCD	
		Untrimmed	Trimmed
Case 1	ε	−1.34%	−0.69%
Case 2	ε + 3°	−1.80%	−0.93%
Case 3	ε + 5°	−1.98%	−0.82%
Case 4	ε + 10°	−1.60%	−0.31%

CL, CD, Cm of the whole aircraft. The trimmed CD with ailerons downward deflection can be calculated based on the following formula:

$$Cm_{trim} = Cm_0 + (\Delta CL_{HT} \cdot cos\alpha + \Delta CD_{HT} \cdot sin\alpha) \cdot \frac{l_{HT}}{C_A} = 0; \tag{3}$$

$$CL_{trim} = CL_0 + \Delta CL_{HT}; \tag{4}$$

$$CD_{trim} = CD_0 + \Delta CD_{HT}. \tag{5}$$

The subscript "trim" denotes the trimmed aerodynamic coefficient of the aircraft, and the subscript "0" denotes the aerodynamic coefficient when the horizontal tail deflection angle is zero. ΔCLHT and ΔCDHT represent the changes in the CL and CD caused by the horizontal tail deflection, respectively. l_{HT} is the lever arm of the horizontal tail. c_A is the reference length used in calculating Cm, which is usually the mean aerodynamic chord length (MAC).

Trimmed ΔCD values are also listed in Table 2. It can be seen that trim has a significant impact on drag reduction. Without considering trim, Case 3 has the best drag reduction effect. While considering trim, the CD reduction values are only about half of the former, and Case 2 has the best effect.

Figure 7 shows the lift distribution of wing along spanwise at CL = CL_{target} for different ailerons deflection angles. The dashed line represents the ideal elliptical distribution. As the ailerons deflection angle gradually increases, the load on the outer wing area increases and approaches the elliptical distribution. However, when the ailerons deflection angle is too large (Case 4), the load on the outer wing area has deviated from the elliptical distribution and can no longer reduce induced drag. In addition, the larger the ailerons deflection angle, the greater the loss of lateral and directional control ability, so the ailerons deflection angle should be reasonably selected.

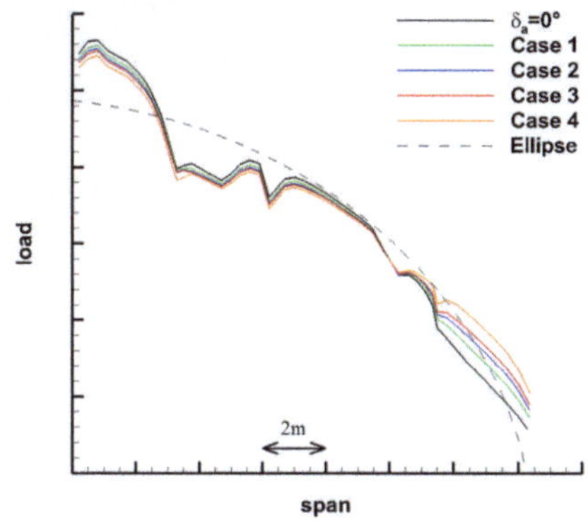

Fig. 7. CL = CL_{target}, lift distribution along spanwise corresponding to different ailerons deflections

3.3 Impact of Gaps at the Ends of Ailerons

For a real aircraft, there are gaps between the aileron and the fixed wing surface. When ailerons are in neutral position, sealing strips on the leading edge and both sides of the ailerons can seal these gaps. When ailerons are deflected, the leading edge sealing strip can still function, but scissor-shaped gaps are formed between both sides of ailerons and the fixed wing surface. Gaps on both sides will reduce the load on the ailerons, directly affecting the drag reduction effect. Therefore, in the design stage, the impact of these gaps must be considered.

Based on the configuration shown in Fig. 5, gaps between the ailerons and the fixed wing surface are cut out, then generate structured grid (Fig. 8). After cutting gaps at the end faces of ailerons, aerodynamic coefficients of different ailerons deflection angles are compared with the base configuration ($\delta a = 0°$, gapless), as shown in Fig. 9. ΔCD values before and after trim are shown in Table 3. Obviously, two gaps of each aileron have a significant impact on the CD reduction. Considering the impact of gaps on trimmed CD, optimal CD reduction is still achieved when the deflection angle is close to Case 2.

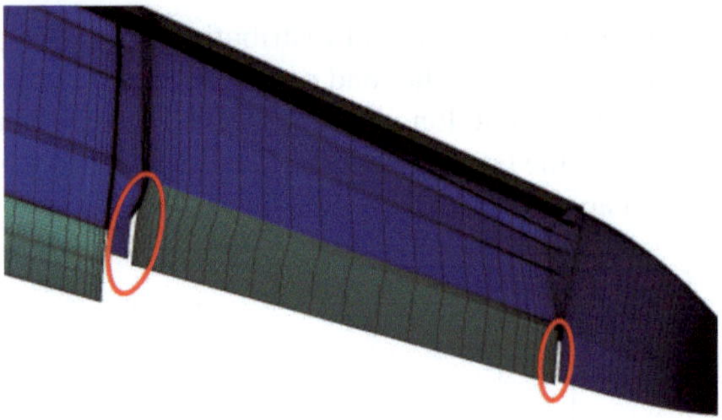

Fig. 8. Grid in aileron area (gaps between the aileron and the fixed wing surface)

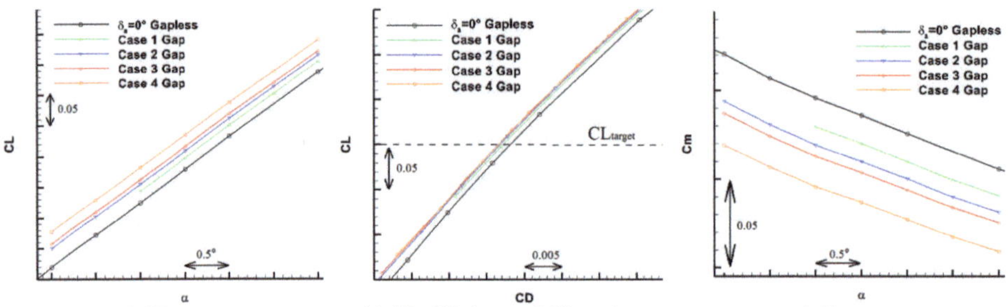

a) CL-α b) CL-CD (around CL_{target}) c) Cm-α

Fig. 9. Aerodynamic coefficients of different ailerons deflection angles (considering impact of gaps)

Table 3. CL = CL_{target}, ΔCD corresponding to different ailerons downward deflection angles (gapped configuration)

No.	δ_a	ΔCD	
		Untrimmed	Trimmed
Case 1	ε	−0.98%	−0.42%
Case 2	ε + 3°	−1.37%	−0.49%
Case 3	ε + 5°	−1.58%	−0.50%
Case 4	ε + 10°	−1.23%	+0.01%

4 Validation Experiments

4.1 Wind Tunnel and Experimental Model

To validate the CFD results, a series of wind tunnel experiments were conducted. The wind tunnel is a low-speed facility operating at atmospheric pressure. The cross-section size of the test section is 8 m × 6 m, which makes it the largest low-speed wind tunnel in

China. The wind speed range is 20 m/s ~ 100 m/s. The Mach number of the experiments was 0.2.

Experimental model was an entire configuration, with slats, flaps, and ailerons can be adjusted to required deflection angles, as shown in Fig. 10. When flaps and ailerons are in neutral positions, the gaps between them and the fixed wing surface were sealed with tapes (Fig. 11). When these components were deflected, scissor-shaped gaps were retained to closely simulate the real aircraft's takeoff configuration. During the operation of the wind tunnel, continuous variable angle of attack was used to obtain complete aerodynamic force data from small angle of attack to the stall section.

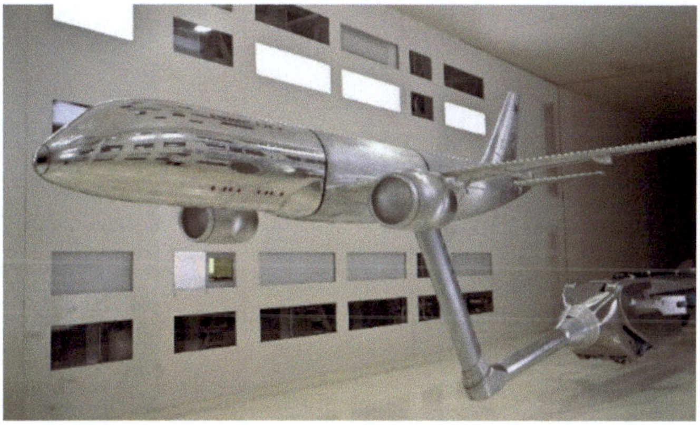

Fig. 10. Experimental model and wind tunnel

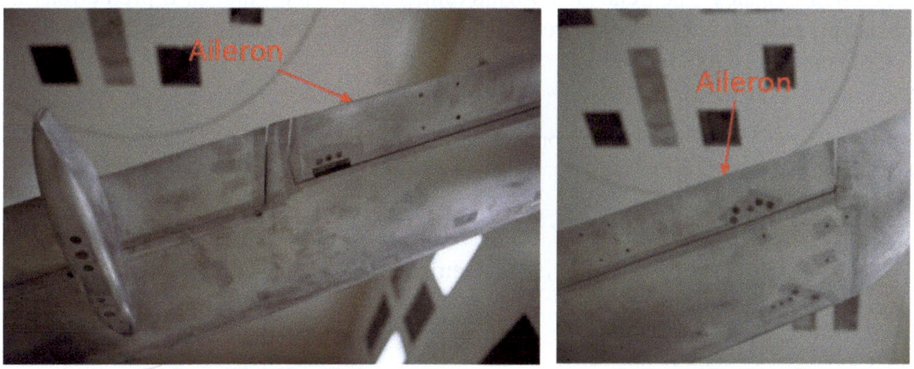

a) Inner side of the aileron b) Outer side of the aileron

Fig. 11. The model's aileron area (gaps were sealed by tapes)

4.2 Experimental Results

The force measurement wind tunnel experiments were conducted separately for the configuration with $\delta_a = 0°$ and Case 1–4. Experimental results are shown in Fig. 12. In the linear region, ailerons downward deflections lead to an increase in CL by about 0.010 – 0.035. In the stall region, CL_{max} is increased by 0.003–0.018, and the shape of C_m-α curve does not change significantly, proving that appropriate ailerons downward deflection angles do not deteriorate the stall characteristics of the aircraft.

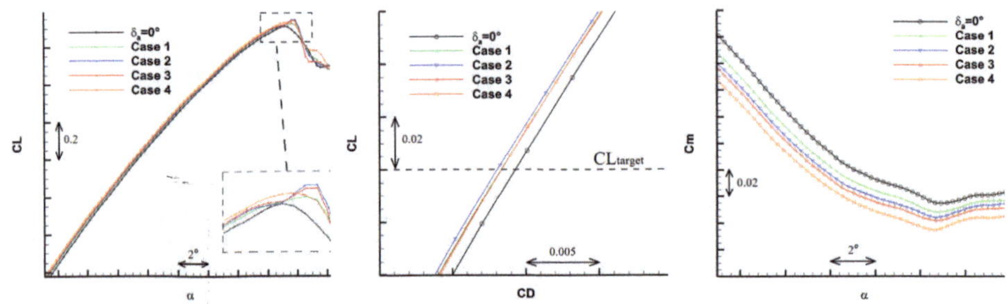

a) CL-α b) CL-CD (around CL_{target}) c) Cm-α

Fig. 12. Experimental results of different ailerons deflection angles

Table 4 presents the experimental values of ΔCD of each aileron deflection at CL = CL_{target} before and after trim, with Case 2 still achieving the most reduction in aerodynamic drag. The deflection angle of Case 4 will lead to an increased in trimmed CD. The correlation between ΔCD and ailerons deflection angle for each configuration is shown in Fig. 13. As the deflection angle increases, the drag reduction benefit initially rises and then diminishes. The trends from wind tunnel experiments and CFD predictions are consistent, and the values (compared with Table 3) are close. The design results based on CFD have been validated.

Table 4. CL = CL_{target}, ΔCD corresponding to different ailerons deflection angles in wind tunnel experiments

No.	δ_a	ΔCD	
		Untrimmed	Trimmed
Case 1	ε	−1.02%	−0.60%
Case 2	ε + 3°	−1.36%	−0.69%
Case 3	ε + 5°	−1.18%	−0.42%
Case 4	ε + 10°	−1.09%	+0.02%

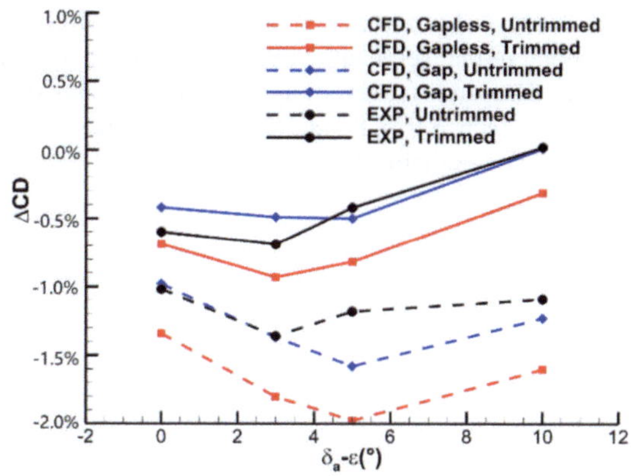

Fig. 13. Curves of ΔCD versus ailerons deflection angle in different configurations, $CL = CL_{target}$

5 Conclusions

For the takeoff configuration of a certain type of civil aircraft, with the goal of improving the lift-to-drag ratio, this paper completed the aerodynamic design of the ailerons downward deflection based on CFD, and conducted wind tunnel experiments for validation. The following conclusions can be drawn:

1. During the takeoff phase, deflecting ailerons downward to a small angle can reduce drag and improve the lift-to-drag ratio without deteriorating the stall performance. But excessive ailerons deflection will result in a lower lift-to-drag ratio.
2. It is reliable to determine the ailerons deflection angle by CFD results.
3. Ailerons downward deflection will cause a significant change in pitch moment. When selecting ailerons deflection angle, the effect of trim on drag must be considered.
4. Gaps will reduce the drag reduction benefits of ailerons deflection downwards. Improving the sealing performance or minimizing gaps width can increase the benefits.

References

1. Anderson, J.D.: Fundamentals of Aerodynamics, 5th ed., pp. 430–434. Aviation Industry Press, Beijing (2014)
2. Bolonkin, A., Giluard, G.B.: Estimated benefits of variable-geometry wing camber control for transport aircraft. NASA-TM-1999–206586. NASA, Washington, DC (1999)
3. Lyu, Z.: High-fidelity aerodynamic design optimization of aircraft configurations. Ph. D. dissertation, pp. 104–122. Ann Arbor: University of Michigan, Ann Arbor, MI, USA (2014)
4. Lyu, Z., Martins, J.R.R.: Aerodynamic shape optimization of an adaptive morphing trailing edge wing. J. Aircr. **52**(6), 1951–1970 (2015)
5. Guo, T., Bai, J., Li, L., Chen, S.: The morphing trailing-edge wing optimization design of the civil aircraft. Sci Sin Tech **48**(1), 55–66 (2018). (in Chinese)
6. Niu, W., Zhang, Y., Chen, H., Zhang, M.: Numerical study of a supercritical airfoil/wing with variable-camber technology. Chin. J. Aeronaut. **33**(7), 1850–1866 (2020)

7. Strüber, H.: The aerodynamic design of the A350 XWB- 900 high lift system. In: 29st Congress of the International Council of the Aeronautical Sciences, Petersburg, Russia (2014)
8. Liu, C., Liu, P., Mao, J.: Research on aerodynamic characteristics of AHDF civil aviation aircraft. Aeronaut. Comput. Tech. **51**(5), 84–86 (2021). (in Chinese)
9. Xie, G.: Aileron droop function implemented by civil aircraft flight control system. Civil Aircr. Des. Res. **4**(8), 18–20 (2011). (in Chinese)

System Control and Mechanical Performance Analysis in Aerospace Engineering

Comparison of Response Time for Pressure of Pintle Solid Rocket Motor Under Different Controllers

Ming Zhu, Wei-Xuan Li(✉), Ying-Kun Li, Xiong Chen, and Hai-Feng Xue

School of Mechanical Engineering, Key Laboratory of Special Engine Technology, Ministry of Education, Nanjing University of Science and Technology, Nanjing 210094, Jiangsu, China
{Zhuming,liweixuan-203,liyingkun,chenxiong}@njust.edu.cn

Abstract. A robust control system incorporating parameter adaptation is developed to mitigate the nonlinearity and parameter uncertainty in pressure control systems of pintle solid rocket motors. The mathematical correlation between pressure and pintle displacement is established through the zero-dimensional interior ballistic model, while the state space equation is derived by integrating the effects of motor inertia load dynamics. Develop a nonlinear control system utilizing robust system and parameter adaptive. By means of simulation, the control performances of traditional PID controller, robust controller, and adaptive robust controller were compared, revealing that the adaptive robust controller exhibited superior control effectiveness.

Keywords: Nonlinearity · Parameter uncertainty · Adaptive robust controller · Traditional PID · Robust controller

1 Introduction

Solid rocket motors (SRM) feature a straightforward design, high operational reliability, ease of maintenance and operation, as well as robust rapid response capabilities. They find extensive applications in aerospace launches, missile systems, as well as orbit and attitude control propulsion. The advancement of military and aerospace technologies necessitates increasingly stringent demands on solid rocket motor technology [1, 2]. The challenges associated with intricate flow control, fixed thrust, and challenging adjustments constrain its capacity to enhance missile range, maneuverability, and combat effectiveness [3].

Given the aforementioned limitations, recent years have witnessed significant research efforts by scholars worldwide, involving experimental investigations, numerical modeling, and related methodologies. Hyun Ko [4] performed both experimental and computational simulations, demonstrating that increased free volume within the combustion chamber under a fixed mass flow rate results in more evident pressure hysteresis effects. Meanwhile, Li [5] introduced a computational approach for determining the nozzle's equivalent throat area. By correlating pintle displacement with the nozzle's effective throat area, a steady-state flow field analysis was conducted, revealing

T. Zielinska (Ed.): PEAE 2025, LNNS 1679, pp. 45–53, 2026.
https://doi.org/10.1007/978-3-032-08221-3_4

that sluggish pressure response was mainly influenced by the high-pressure exponent and the inherent characteristics of the low-temperature propellant. Addressing the issue of delayed pressure response leading to thrust adjustment lag in pintle-controlled solid rocket motors, Miao formulated a transient model for such systems and determined the optimal pintle configuration through thrust regulation and specific impulse efficiency assessments. Simulations further suggested that pressure response behavior is strongly linked to propellant combustion rate attributes and pintle movement dynamics, with a noticeable thrust delay occurring at pintle speeds surpassing 2 m/s.

Although pintle displacement enables direct and precise control over combustion pressure and thrust output, a misalignment often arises between the actual pintle position and its ideal theoretical placement for optimal thrust performance. If the pintle's displacement at the nozzle throat's limit position deviates even slightly, the resulting pressure may exceed the structural tolerance of the nozzle, potentially causing catastrophic failure [6]. To address this challenge, one viable solution involves refining controller optimization algorithms to improve the accuracy of pressure and thrust regulation in pintle-controlled solid rocket motors. Among established control strategies, the PID (Proportional-Integral-Derivative) algorithm remains a widely recognized conventional approach. Building upon combustion chamber dynamics, Cha [6] designed specialized control systems for pressure and thrust modulation, employing three distinct algorithmic methods. Comparative simulations revealed that feedback linearization control outperformed others in pressure stabilization, whereas fuzzy PID control demonstrated greater efficacy in thrust adjustment.

In this paper, an adaptive robust controller is employed to improve the precision, resilience, and stability of the control system, effectively addressing the problem of delayed pressure response caused by inaccuracies in pintle position. By comparing the designed controller with traditional PID and robust controllers, it was observed that this controller demonstrates excellent performance in terms of accuracy, robustness, and stability.

2 Mathematical Model

2.1 Dynamic Models of Combustion Chamber Pressure

Figure 1 presents a schematic representation of the simplified pintle nozzle configuration for a solid rocket motor employing end-burning propellant combustion. The system maintains constant initial throat and nozzle exit areas, with pintle displacement serving as the primary mechanism for throat area modulation. This adjustment directly influences combustion chamber pressure dynamics, thereby enabling thrust control. The pressure transient within the combustion chamber can be mathematically described by the following zero-dimensional interior ballistic equation [6]:

$$\begin{cases} \dfrac{d(P_c)}{dt} = \dfrac{\Gamma^2 C^{*2}}{V_c}(\rho_P A_b v P_c^n - \rho_c A_b v P_c^n - \dfrac{P_c A_t}{C^*}) \\ \dfrac{d(\rho_c)}{dt} = \dfrac{1}{V_c}(\rho_P A_b v P_c^n - \rho_c A_b v P_c^n - \dfrac{P_c A_t}{C^*}) \\ \dfrac{d(V_c)}{dt} = A_b v P_c^n \end{cases} \tag{1}$$

$$\Gamma = \sqrt{\varepsilon}\left(\frac{2}{\varepsilon + 1}\right)^{\frac{\varepsilon+1}{2(\varepsilon-1)}} \tag{2}$$

In (1), P_c corresponds to the steady-state combustion chamber pressure. The system parameters include: propellant density (ρ_P) of 1700 kg/m^3, characteristic velocity (C^*) of 1500 m/s, and combustion rate coefficient (v) of 6.9 × 10^{-6}. Additional parameters comprise the burning surface area (A_b) of 0.017671 m^2, equivalent throat area (A_t), and the pressure index (n) of 0.5. The gas density (ρc), combustion chamber free volume (V_c) and specific heat ratio ($\varepsilon = 1.25$) complete the thermodynamic characterization of the system [7].

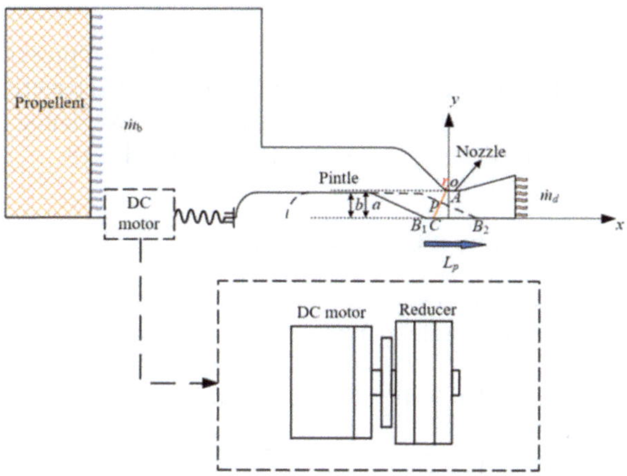

Fig. 1. Simplified model of pintle-nozzle and motor.

2.2 Mathematical Model of Actuator

The electromechanical actuator system employed in this study incorporates a permanent-magnet DC motor, which converts electrical energy from a direct current source into precise mechanical displacement for pintle position control. The governing equations for the inertial load dynamics are expressed as follows [8]:

$$J\ddot{y} = ku - B\dot{y} - F_f + \Delta(t) \tag{3}$$

where J represents the inertial load. y, $\dot{y}$, and $\ddot{y}$ denote the angular displacement, angular velocity, and angular acceleration of the motor, respectively. k is the torque constant, u represents the controller input, B is the coefficient of viscous friction, and F_f denotes the total friction moment. $\Delta(t)$ represents other system disturbances.

3 Adaptive Robust Controller Design

3.1 State Space Equation

The establishment of the pressure state space equation serves as the basis for designing control inputs. By integrating the dynamic pressure equation with the mathematical model of the actuator, the pressure state space equation is derived as [9]:

$$\begin{cases} \dot{x}_1 = \dfrac{1}{V_c}(\theta_1 x_1^n - \theta_2 \rho_c x_1^n - \theta_3 x_1 x_2) \\ \dot{x}_2 = g(t) x_3 \\ \dot{x}_3 = \theta_4 u - \theta_5 x_3 - \theta_6 F_f(x_3) + \Delta_1(t) \end{cases} \tag{4}$$

In (4), $\theta_4 = k/J$, $\theta_5 = B/J$, $\theta_6 = 1/J$ and $\Delta_1(\mathrm{t}) = \Delta(\mathrm{t})/J$. $\Delta_1(\mathrm{t})$ represents perturbations, while F_f incorporates unmodeled errors, resulting in unmodeled uncertainties within the system. $\theta_1 = \Gamma^2 C^{*2} \rho_P A_b v$, $\theta_2 = \Gamma^2 C^* 2 A_b v$, and $\theta_3 = \Gamma^2 C^*$. We define the state variable as $x = [x_1, x_2, x_3]^{\mathrm{T}}$, where $x_1 = P_c$, $x_2 = A_t$, and $x_3 = L_P/\mathrm{d}t$ [9].

3.2 Design of Adaptive Robust Controller Input

Parameter adaptation is the process of guaranteeing tracking performance of a system under conditions of uncertain model parameters. Hence, it is defined as:

$$\begin{cases} \theta = [\theta_1, \theta_2, \theta_3, \theta_4, \theta_5, \theta_6]^T \\ \hat{\theta} = \left[\hat{\theta}_1, \hat{\theta}_2, \hat{\theta}_3, \hat{\theta}_4, \hat{\theta}_5, \hat{\theta}_6\right]^T \\ \tilde{\theta} = \left[\tilde{\theta}_1, \tilde{\theta}_2, \tilde{\theta}_3, \tilde{\theta}_4, \tilde{\theta}_5, \tilde{\theta}_6\right]^T \end{cases} \tag{5}$$

where θ represents the true value of the parameter vector, $\hat{\theta}$ is the estimate of the parameter vector, $\tilde{\theta}$ is the error value of the parameter vector, and $\tilde{\theta} = \hat{\theta} - \theta$. By combining (5) the error rate is:

$$\begin{cases} \dot{e}_1 = -\dfrac{1}{V_c}\tilde{\theta}_1 x_1^n + \dfrac{1}{V_c}\tilde{\theta}_2 \rho_c x_1^n + \dfrac{1}{V_c}\tilde{\theta}_3 x_1 \alpha_1 - \dfrac{1}{V_c}\theta_3 x_1 e_2 - k_1 e_1 \\ \dot{e}_2 = g(t) e_3 - k_2 e_2 \\ \dot{e}_3 = \tilde{\theta}_5 x_3 + \tilde{\theta}_6 F_f - k_3 e_3 + \Delta_1(t) - \tilde{\theta}_4 u \end{cases} \tag{6}$$

So as to obtain the controller input [9]:

$$u = \frac{1}{\hat{\theta}_4}\left[\hat{\theta}_5 x_3 + \hat{\theta}_6 F_f + \dot{\alpha}_2\right] - \frac{1}{\hat{\theta}_4} k_3 e_3 \tag{7}$$

Table 1. True, estimated, and error values of parameters

Parameter	True values	Estimates values	|error values|
θ_1	201.966	201.4–201.5	0.466–0.566
θ_2	0.1188	0.12–0.14	0.0012–0.0212
θ_3	649.574	647.5–648.5	1.074–2.074
θ_4	0.2	0.2493–0.25	0.0493–0.05
θ_5	0.02	0.025–0.02500000014	0.00500000014–0.005
θ_6	0.2	0.249999999975–0.25	0.049999999975–0.05

4 Simulation Results

4.1 Parameter Settings

The system's dynamic model incorporates the following inertial and electromechanical parameters: moment of inertia $J = 5$ kg·m^2, torque constant $k = 1$ N·m/v, and viscous damping coefficient $B = 0.1$ N·s/m. The three control strategies under investigation are parameterized as follows.

1) PID controller (hereinafter referred to as PID for brevity)

The PID control architecture employed in this system implements a conventional three-loop cascade structure encompassing position, velocity, and current regulation. This control scheme combines proportional (P), integral (I), and derivative (D) actions to achieve system stabilization. The controller gains are initially configured at $k_p = 50$, $k_i = 50$, and $k_d = 50$, with provisions for dynamic adjustment to accommodate varying operational requirements and optimize system performance.

2) Robust controller (hereinafter referred to as RC for brevity)

A robust control scheme maintains closed-loop system stability through appropriate feedback gain selection. Considering the rapid pressure transients occurring within millisecond timescales, careful determination of feedback gains becomes critical for stability preservation. The controller architecture employs the following gain parameters: $k_1 = 1000$, $k_2 = 1000$, $k_3 = 2000$. Structural uncertainties are characterized by the parameter vector $\theta = [\theta_4, \theta_5, \theta_6]^{\mathrm{T}} = [0.2, 0.02, 0.2]^{\mathrm{T}}$, representing system modeling variations.

3) Adaptive robust controller (hereinafter referred to as ARC for brevity)

An adaptive robust control scheme ensures system stability through real-time parameter error estimation. Table 1 presents the comparative analysis of actual parameters, their estimated values, and corresponding estimation errors during the adaptation process. Based on these empirical results, the following conclusions can be drawn:

$$\left[\dot{\hat{\theta}}_1, \dot{\hat{\theta}}_2, \dot{\hat{\theta}}_3, \dot{\hat{\theta}}_4, \dot{\hat{\theta}}_5, \dot{\hat{\theta}}_6\right]^{\mathrm{T}} = [\tau_1\varphi_1, \tau_2\varphi_2, \tau_3\varphi_3, \tau_4\varphi_4, \tau_5\varphi_5, \tau_6\varphi_6]^{\mathrm{T}} \tag{8}$$

The parameters $\Gamma = \text{diag}(\tau1, \tau2, \tau3, \tau4, \tau5, \tau6)$ are set to $\Gamma = \text{diag}(10^{-10}, 10^{-10}, 10^{-10}, 10^{-10}, 10^{-10}, 10^{-10})$.

Figure 2 illustrates the desired pressure curve, which is smooth and continuously differentiable up to the third degree. The simulation lasted for a total of 3 s, with a step size of 1×10^{-5}. The pressure rose from 0.2 s to 0.3 s over a duration of 0.1 s.

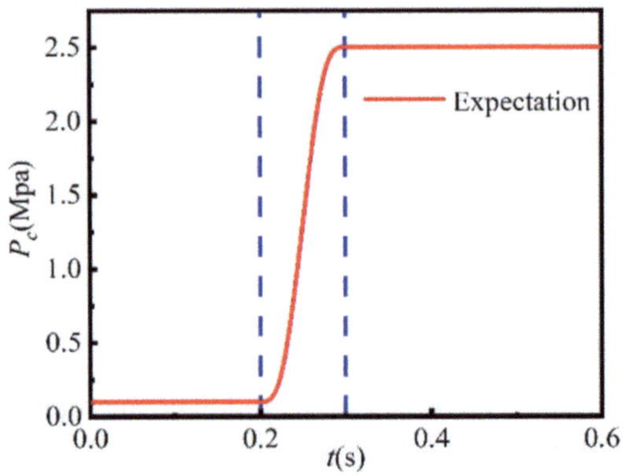

Fig. 2. Pressure expectation curve.

4.2 Comparison of Three Controller Simulations

The control inputs applied to the physical system consist of analog voltage signals, whose operational ranges are bounded within [−10, 10] volts due to hardware constraints, as implemented across all three control schemes. Figure 3 presents the comparative performance of the PID, robust controller (RC), and adaptive robust controller (ARC) input signals. As shown in Fig. 3(a), the ARC generates control commands with substantially greater amplitude prior to saturation compared to alternative control approaches. Figure 3(b) reveals that the PID controller produces highly oscillatory inputs upon reaching saturation limits. While the RC initially exhibits unstable behavior, its control signals progressively converge to stable values. In contrast, the ARC maintains consistent stability throughout the entire operational period. This comparative analysis of controller inputs demonstrates that the ARC achieves superior control performance, particularly in maintaining system stability under input constraints.

Three distinct free volume configurations (3.5×10^{-6} m^3, 2.4×10^{-4} m^3, 2.4×10^{-3} m^3) were selected to evaluate controller efficacy across different operational scales. Figure 4 demonstrates that both the adaptive robust controller (ARC) and robust controller (RC) exhibit significantly greater pressure rate variations compared to the PID controller at all volume levels. However, comparative analysis reveals that at intermediate and larger free volumes (2.4×10^{-4} m^3 and 2.4×10^{-3} m^3), the pressure response characteristics of ARC and RC converge, suggesting the need for detailed temporal analysis when assessing controller performance. This observation underscores the necessity of incorporating response time metrics in comprehensive controller evaluations.

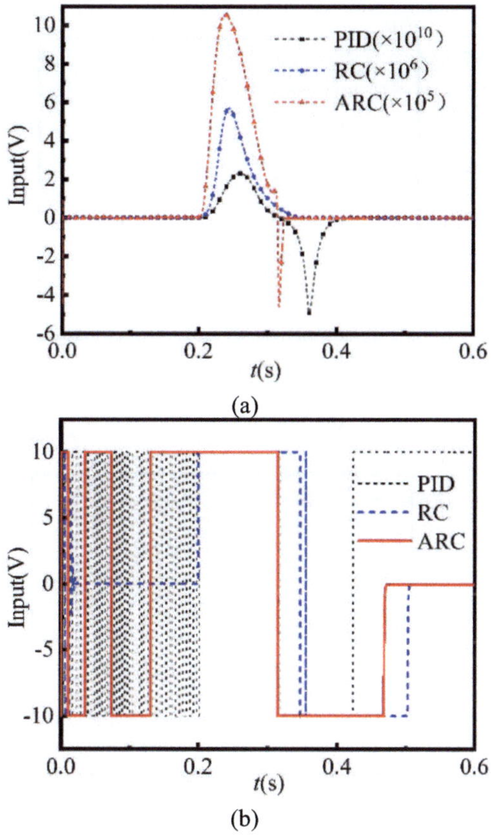

Fig. 3. Control inputs of three controllers before and after limitation.

Table 2 presents the temporal response characteristics of the three control systems under varying free volume conditions. The adaptive robust controller (ARC) consistently achieves superior response speeds, demonstrating performance improvements of 10.01% and 13.16% (small volume), 0.03% and 1.54% (moderate volume), and 0.06% and 22.39% (large volume) relative to the robust controller (RC) and PID controller, respectively. Notably, controller performance differences become statistically insignificant at intermediate volume levels, while exhibiting marked variation at smaller volumes. These findings suggest that optimal controller operation occurs under moderate free volume conditions, where all three control architectures demonstrate comparable response characteristics.

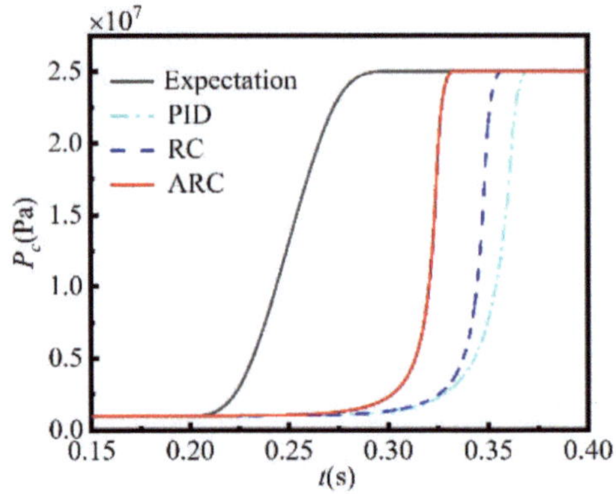

(a) $V_{c0}=3.5\times10^{-6}m^3$

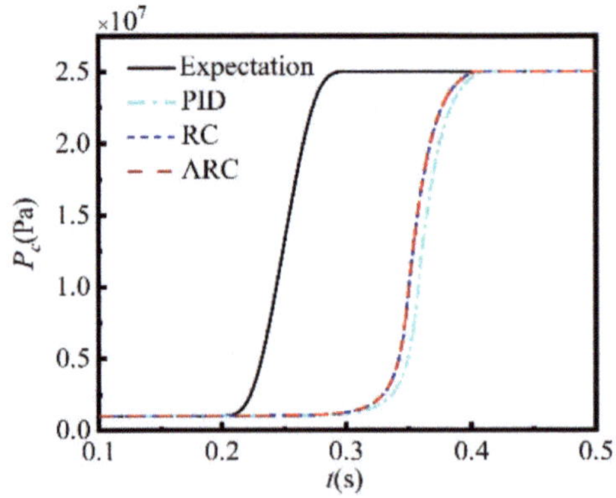

(b) $V_{c0}=2.4\times10^{-4}m^3$

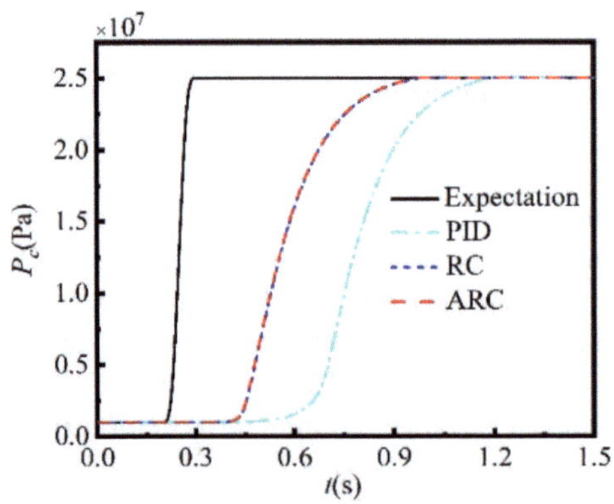

(c) $V_{c0}=2.4\times10^{-3}m^3$

Fig. 4. Pressure Response Comparison Chart.

Table 2. Comparison of Pressure Response Times for Different Controllers

	Controller	3.5×10^{-6} m^3	2.4×10^{-4} m^3	2.4×10^{-3} m^3
Response time/s	**PID**	0.36507	0.40889	1.20037
	RC	0.35493	0.40282	0.98140
	ARC	0.32262	0.40269	0.98078

5 Conclusion

This study investigates the nonlinear characteristics and parametric uncertainties inherent in pintle-type solid rocket motor systems. A novel adaptive robust control (ARC) scheme was developed through integration of the combustion chamber pressure dynamics with the electromechanical actuator's inertial load model. Comparative simulations reveal the ARC framework consistently outperforms both robust control (RC) and conventional PID architectures across the entire operational range of free volume conditions. Notably, controller performance differentials become marginal under intermediate volume configurations. Quantitative analysis demonstrates the ARC system achieves temporal response improvements of 0.03% relative to RC and 1.54% compared to PID control at medium free volume conditions.

References

1. Bao, L.R., Chen, Y.Y., Chen, S.H., Zhang, W., Wu, L.Z., Shen, R.Q., et al.: Research progress on controllable solid propulsion. J. Propul. Technol. **41**(5), 961–973 (2020)
2. Lv, X., He, G.Q., Liu, P.J., Li, Q.: Requests and suggests for basic research on SRM combustion and flow. J. Astronaut. **40**(10), 1157–1166 (2019)
3. Hou, X., Fu, P., Wu, Y.: Energy management technology of SRM and its development. J. Solid Rocket Technol. **40**(1), 1–6+23 (2017)
4. Ko, H., Lee, J.H., Chang, H.B., Yoon, W.S.: Cold tests and the dynamic characteristics of the pintle type solid rocket motor. In: 49th AIAA/ASME/SAE/ASEE Joint Propulsion Conference. California (2013)
5. Li, J., Li, J., Wang, Y.L., Qin, F.: Study on performance of pintle controlled thrust solid rocket motor. J. Solid Rocket Technol. **30**(6), 505–509 (2007)
6. Cha, J., de Oliveira, É.J.: Performance comparison of control strategies for a variable-thrust solid-propellant rocket motor. Aerospace **9**(6), 325 (2022)
7. Deng, H., Li, Z.H., Zhang, S.K., Yang, M.J., Li, W.P., Yan, O.P.: Pintle profile design method based on zero-dimensional interior ballistic model and performance for variable thrust motor. J. Propul. Technol. **43**(11), 34–43 (2022)
8. Xu, Z.B., Yao, J.Y., Ma, D.W., Yang, G.C.: Robust control of DC motors based on disturbance estimation. In: 34th Chinese Control Conference. Hangzhou (2015)
9. Zhu, M., Li, Y.K., Chen, X., Mi, J.J., Li, W.X., Xue, H.F.: Design of nonlinear control method for pressure of pintle solid rocket motor. Acta Astronautic **215**, 296–307 (2024)

Experimental Study on the Spatial Characteristics of Hypersonic Gap-Compression Ramp Shock Wave Boundary/Layer Interaction

Xiaohui Li, Hongliang Xiong, Hongwei Wang, Zhan Huang(✉), Shaojie Ren, and Weilong Shi

Aerodynamics Research Institute, China Academy of Aerospace Aerodynamics, Beijing, China
xfd_huangzh@sina.com

Abstract. The gap-compression ramp is a simplified shape of the trailing edge rudder of hypersonic vehicle which accompanied with complicated local unsteady separation flow. It has an obvious influence to the aerodynamic performance and affects the control effect of the rudder. Particle image velocimetry and schlieren were used to obtain the local spatial flow characteristics of the gap-compression ramp. The flow characteristics and spatial shock wave distribution of the local interference region were obtained under different rudder gaps. The variation characteristics of the flow separation under different rudder gaps were compared and found that there were suction and overflow effects at the gap. It has an obvious influence on the size and location of separation flow. Besides, the oscillation characteristics of the separation shock were also different to the common compression ramp.

Keywords: wave/boundary layer interaction · gap-compression ramp · separation flow

1 Introduction

Shock wave/boundary layer interaction is a typical flow phenomenon that is widespread in the high speed flow and has significant effects in aerodynamic performance of the aircraft, such as transonic airfoil, high-speed aircraft control surfaces, high-speed engine intakes, turbine blades, and helicopter rotors [1]. Flow separation caused by shock boundary layer interference lead to rapid changes of aerodynamic and aerodynamic thermal loads on the aircraft surface. The instability of the boundary layer flow induces a sharp increase in local heat flow, resulting in deviation from the original design condition of the aircraft. The instability of shock wave will lead to buffeting and aerodynamic fatigue of aircraft [2, 3].

Trailing edge rudder/all-movable rudder and body flap are widely used in high speed aircraft. Andreas Mack et al. [4] studied the influence of flow in hinge gap of body flap on thermal load of X-38 aircraft to explore the interference between body flap gap and body and found that there was strong flow between body flap and fuselage when there

T. Zielinska (Ed.): PEAE 2025, LNNS 1679, pp. 54–63, 2026.
https://doi.org/10.1007/978-3-032-08221-3_5

was a gap. Aiming at the aerodynamic heating problem of all-movable rudder, the US Naval Ground Weapons Center [5, 6] used phase change coating technology to carry out experimental research on the thermal environment of the rudder surface and the interference zone of the cone-rudder shape, and obtained the heat flow distribution of rudder surface, rudder shaft and the interference zone under different height gaps and Reynolds number. It was found that the aerodynamic heating of the rudder shaft and the interference zone of the rudder shaft was significantly increased with the increase of the gaps. At present, the mechanical and thermal characteristics of the crease-compression corner are not fully known, and the experimental research needs to be strengthened.

Aiming at the common shock wave/boundary layer interaction in the trailing edge rudder of high-speed aircraft. PIV and Schlieren measurements of local structural space flow in the gap-compression corner were carried out in this paper, and the flow characteristics and spatial wave system distribution in the local interference region were obtained. The fluorescence oil film friction measurement was carried out to obtain the surface flow map in the local interference region. It provides support for the study of crease-compression corner shock wave-boundary layer interference.

2 Experimental Setup and Measurement Methods

2.1 Wind Tunnel Facility and Experimental Specimen

The experimental environment was provided by a small hypersonic wind tunnel which is driven by high pressure gas and multi-stage injection of medium pressure gas. It can provide Ma = 6.0 inflow environment.

As shown in Fig. 1 and Table 1, the width of the specimen is 100 mm, when there is no gap, the total length is 285 mm, the compression ramp is placed on the surface, and the leading edge of the compression ramp is 205 mm away from the leading edge of the flat. A replaceable transition belt is designed at the front of the specimen, with a height of 0.8 mm and a distance of 40 mm from the model leading edge.

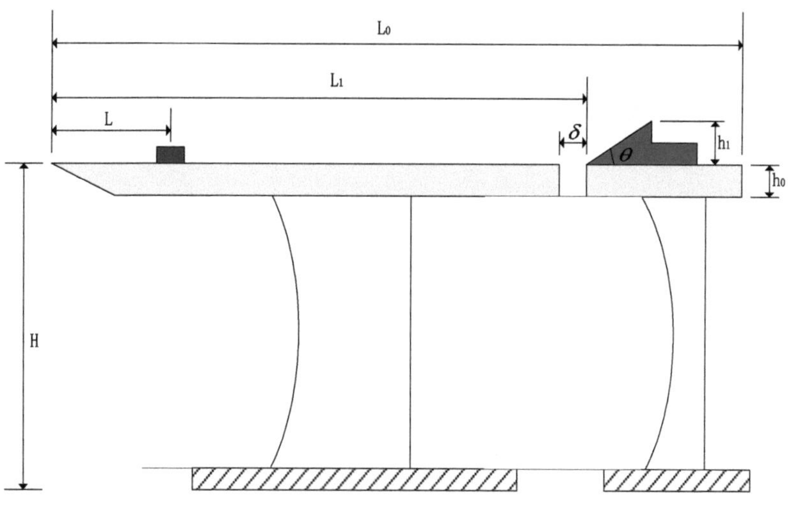

Fig. 1. Experimental specimen

Table 1. Parameter of specimen

Parameter	Description	Value
δ	gap width	1 mm/2 mm/3 mm
L_0	total length of plate	285 + δ mm
L_1	the distance from ramp to model leading edge	205 mm
H	Total height	210 mm
h_0	plate thickness	10 mm
h_1	height of compression ramp	20 mm
θ	Angle of compression ramp	42°
L	distance from transition to the leading edge	40 mm

2.2 Measurement Setup and Experimental Condition

Particle image velocimetry (PIV) is used to measure the spatial flow field structure and velocity field. The specimen adopts the bottom support to measure the x-y plane. As shown in Fig. 2, the laser light is incident from the top window, and the camera shoots from the side window. The spanwise center section (z = 0) of the specimen is measured, and the width of the gap is adjusted by moving the ramp backward. The Schlieren system adopts a traditional knife-edge schlieren instrument and a high-speed camera for flow field. The camera used Photron FASTCAM Nova high-speed camera with a pixel size of 1024 × 1024 pixels with sensitivity of ISO 64000. The camera resolution is 1024 × 1024 pixels.

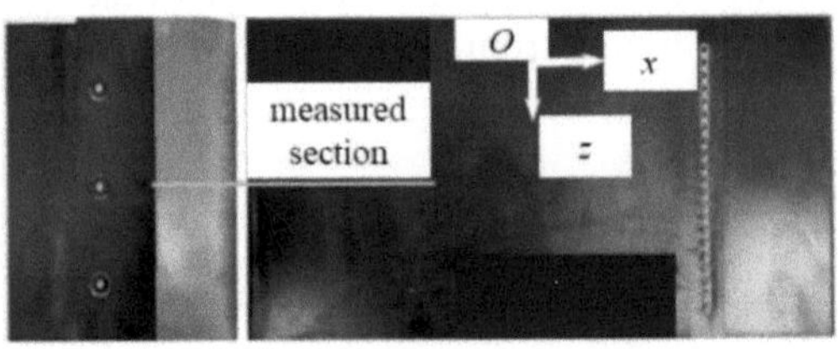

Fig. 2. Measure section of PIV

The Table 2 is the working condition of the experiments. The total incoming pressure is 1.2 Mpa and 2 Mpa, the total incoming temperature is 453K, and the widths of the gaps are 0 mm, 1 mm, 2 mm and 3 mm respectively。

Table 2. Experimental condition

number	Pressure (Mpa)	Gap width(mm)	temperature (K)
1	1.2	0	453
2	1.2	1	
3	1.2	2	
4	1.2	3	
5	2	0	
6	2	1	
7	2	2	
8	2	3	

3 Results and Discuss

3.1 Flow Structure

In order to clearly characterize the position of the shock wave, the gradient sum of squares *DV* of normal velocity in two flow directions are used to identify the position of the shock wave, because the flow direction will be deflected when passing through the shock wave.

$$DV = \sqrt{(\frac{dv}{dx})^2 + (\frac{dv}{dy})^2} \tag{1}$$

Figures 3, 4, 5 and 6 shown the normal gradient cloud image with a gap of 0-3 mm at the condition of total pressure is 1.2 Mpa. It can be seen that under the reverse pressure gradient of the ramp, an obvious separation zone is generated, and the location of the separation point, the reattachment point and the size of the separation zone can be clearly distinguished. With the increase of gap width, the time-mean separation zone decreases gradually.

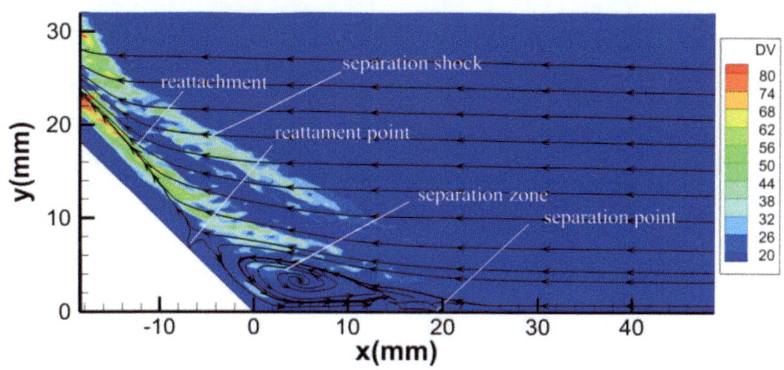

Fig. 3. 1.2 Mpa-0 mm

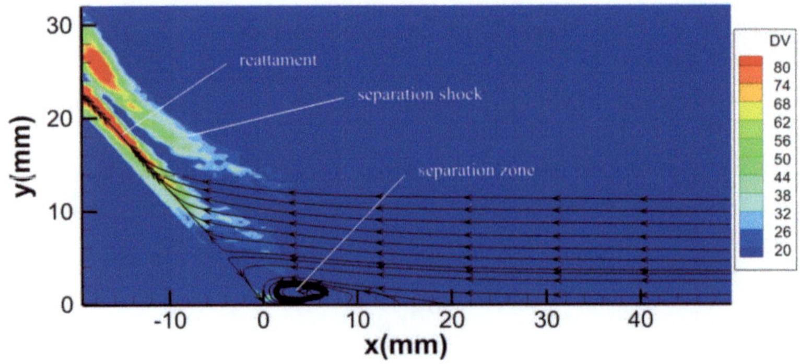

Fig. 4. 1.2 Mpa-1 mm

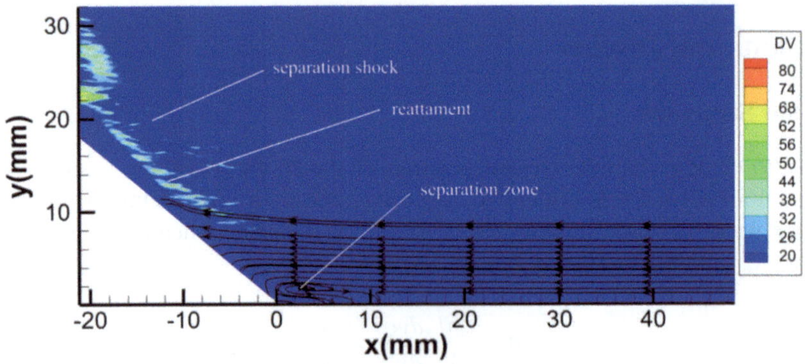

Fig. 5. 1.2 Mpa-2 mm

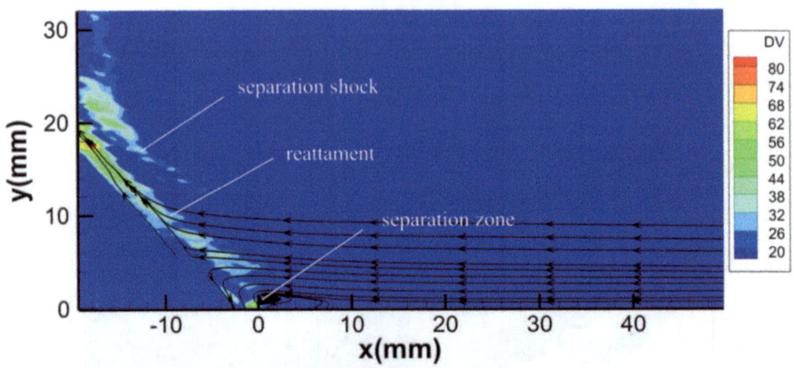

Fig. 6. 1.2 Mpa-3 mm

Figures 7, 8, 9 and 10 shown the normal gradient cloud image with a gap of 0-3 mm at the condition of total pressure is 2.0 Mpa. When the total pressure of the incoming flow increases, the high Reynolds number flow provides stronger kinetic energy. The separation region induced by shock wave becomes larger in the case of the same gap. And the location of the separation point, the reattachment point and the size of the separation zone can be clearly distinguished. With the increase of gap width, the time-mean separation zone decreases gradually.

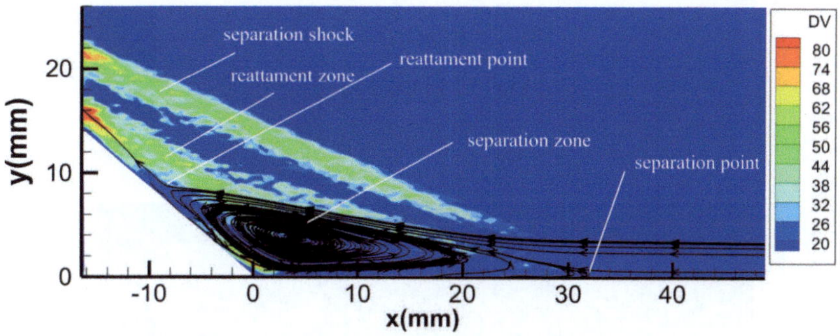

Fig. 7. 2 Mpa-0 mm

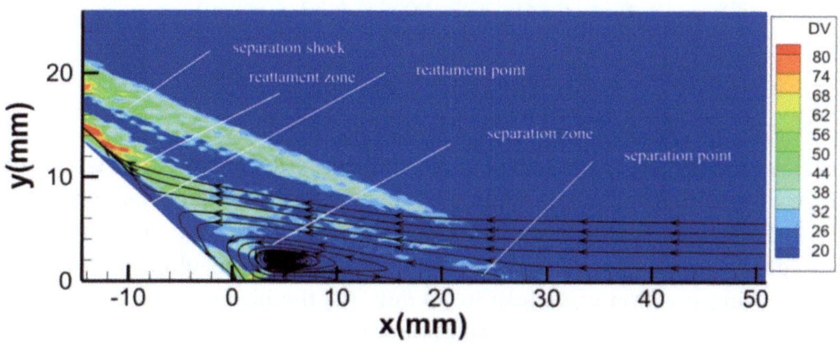

Fig. 8. 2 Mpa-1 mm

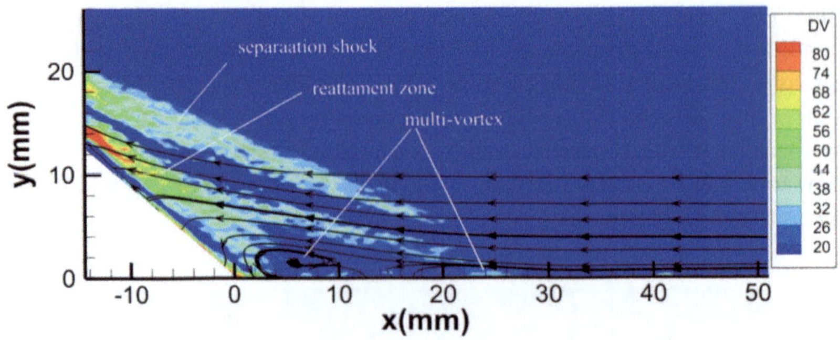

Fig. 9. 2 Mpa-2 mm

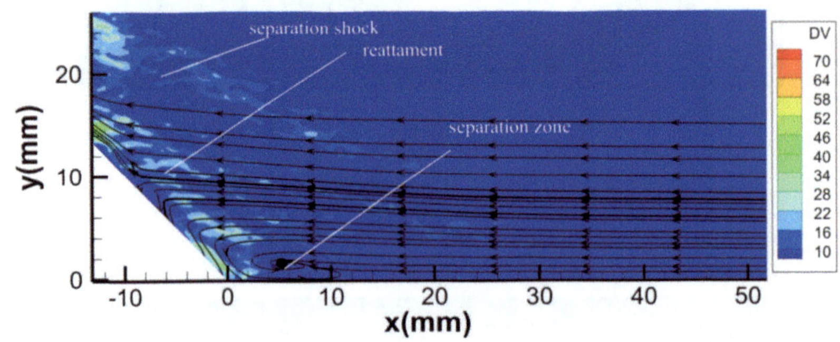

Fig. 10. 2 Mpa-3 mm

In particular, when the gap width is 3 mm, the separation shock wave has a significant wide range of oscillation. As shown in Fig. 11, (a) the instantaneous vortex structure basically disappeared, (b) the instantaneous tiny vortex structure appeared and began to develop gradually, (c) the instantaneous vortex structure developed strongly, and its separation starting position was close to 30 mm, (d) the instantaneous vortex structure almost reached the maximum, and its separation starting position was 34 mm. The results correspond to the Schlieren results shown in Fig. 12.

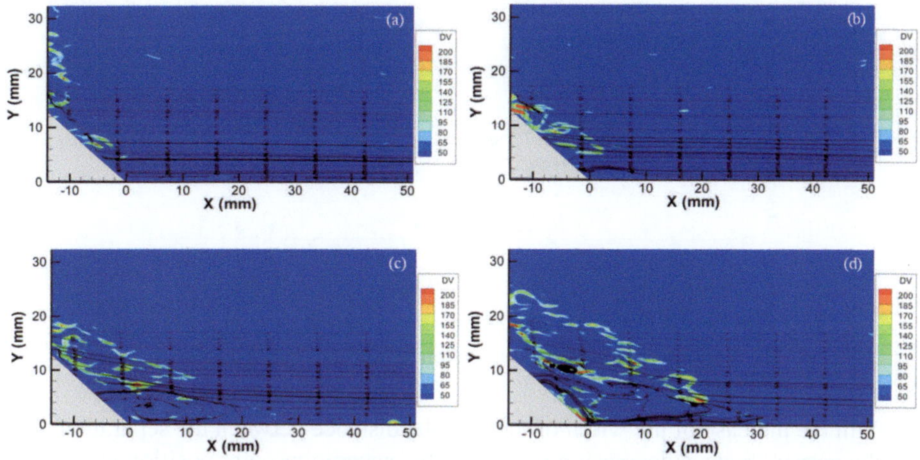

Fig. 11. Normal gradient cloud images at different times

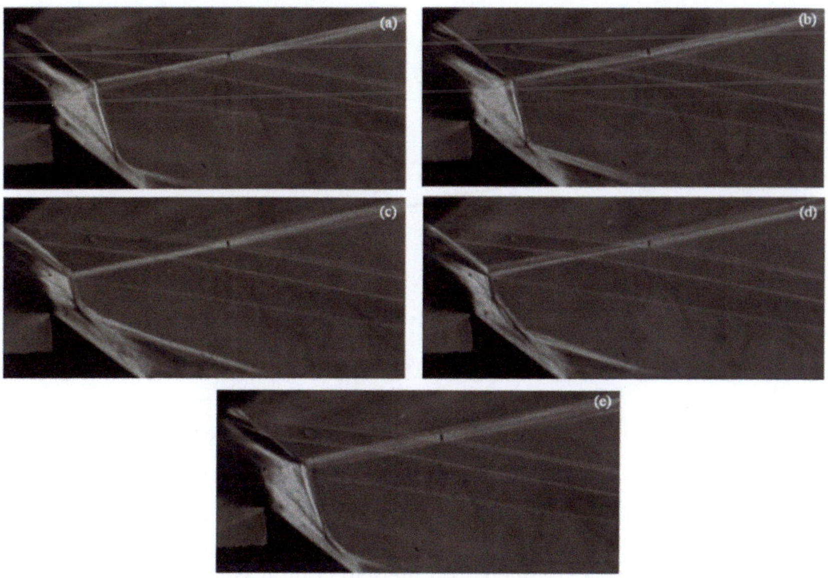

Fig. 12. Schlieren image

3.2 Quantitative Analysis

Figure 13 is the separation shock angle in different cases. When the total pressure is 1.2 Mpa, the separation shock angle is relatively higher than that at 2 Mpa, but the gap width has little influence on the shock angle under the same working condition.

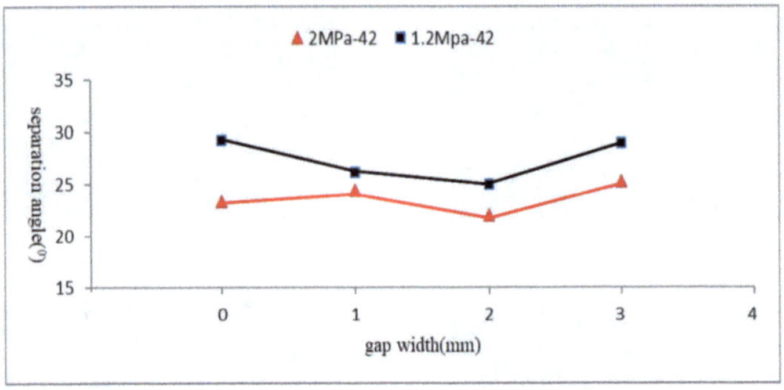

Fig. 13. The angle of separation shock in different cases

With the increase of the width of the gap, the distance between the separation point and the corner shows a decreasing trend, and the separation point position at 2 Mpa is larger than that at 1.2 Mpa, as shown in Fig. 14.

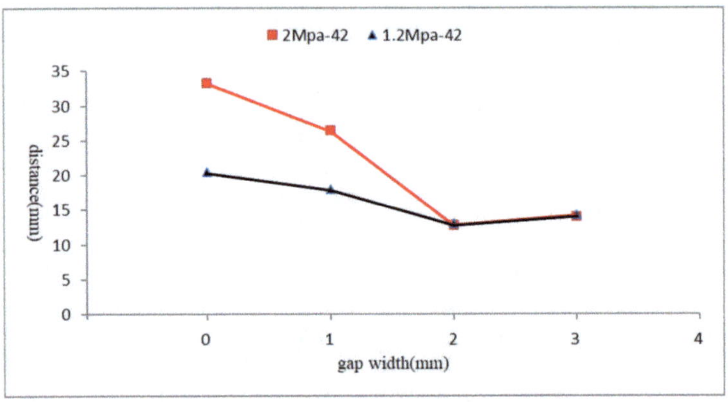

Fig. 14. The distance between the separation point and the leading edge of compression ramp

4 Conclusion

An experimental study was conducted on the gap compression ramp specimen with a compression angle of 42 degrees. The location of the separation zone, the separation point, the reattachment point and the spatial structure distribution of the shock wave were obtained. The following conclusions were given:

(1) When the total pressure changes from 1.2 Mpa to 2 Mpa, the separation zone caused by shock waves becomes larger;
(2) With the increase of gap width, the separation zone gradually decreases;
(3) When the total pressure is 2 Mpa with gap width 3 mm, the shock boundary layer interaction structure shows obvious oscillation characteristics.

References

1. Holden, M.S.: A review of aerothermal problems associated with hypersonic flight. In: 24th AIAA Aerospace Sciences Meeting and Exhibit, AIAA Paper, p. 0267 (1986)
2. Baldwin, A., Mears, L.J., Kumar, R., Alvi, F.S.: Effects of Reynolds number on swept shock-wave/boundary-layer interactions. AIAA J. **59**(10), 3883–3899 (2021)
3. Adler, M.C., Gaitonde, D.V.: Influence of separation structure on the dynamics of shock/turbulent-boundary-layer interactions. Theor. Computer Fluid Dyn. **36**, 303–326 (2022)
4. Mack, A., Schaefer, R.: Fluid structure interaction on a generic body-flap model in hypersonic flow. J. Spacecr. Rocket. **42**(5), 769–779 (2025)
5. Winkelmann, E.: Experimental investigations of a fin protuberance partially immersed in a turbulent boundary layer at Mach 5. AD743903 (1974)
6. Gillerlain, J.D.: Experimental investigation of a fin-cone interference flow field at Mach 5. ADA029999 (1975)

Research on the Influence of Crosswind on the Inlet of Aircraft APU

Jin-Hua Lv[1(✉)], Pan Teng[1], and Hai-Tao Zhang[2]

[1] Shanghai Aircraft Design and Research Institute, Pudong New District, Shanghai 200100, China
2573915726@qq.com, Tengpan@comac.cc, zhanghaitao_hd@caac.gov.cn
[2] Powerplant Division, Shanghai Aircraft Airworthiness Certification Center, Xuhui District, Shanghai 200000, China
zhanghaitao_hd@caac.gov.cn

Abstract. Auxiliary Power Unit (APU) is an important subsystem of modern large and medium-sized passenger aircraft. This paper first used CATIA to construct a model of an aircraft and its APU inlet system, and then used Star CCM+ to simulate the APU inlet flow field. It was found that the ground wind direction has a small impact on the APU inlet total pressure recovery coefficient, but a significant impact on the inlet distortion. The crosswind not only significantly increases the APU inlet distortion, but also changes the direction of the APU inlet distortion. The position of the APU inlet also has an impact on the APU inlet distortion. Under 90° crosswind (the crosswind comes from the aircraft side where the APU inlet is located), the APU inlet distortion is maximum.

Keywords: crosswind · APU · total pressure recovery coefficient · inlet distortion

1 Introduction

Auxiliary power unit is a small power system that can independently output compressed air and electricity in addition to the main engine on modern large and medium-sized aircraft [1]. It is mainly used to continuously supply compressed air and electricity to the aircraft without relying on ground power equipment when the main engine stops working [2]. The inlet system is an important subsystem of APU [3, 4]. Its main function is to provide the air required by the APU under various working conditions [5]. The inlet system, at the forefront of the APU system cycle, has a significant impact on the normal operation and performance of the APU entire system, and is a precondition for the normal operation of APU other components [6].

In recent years, research on the APU inlet system has become one of the hot research issues. Ralston et al. designed an aircraft APU air intake with ram and flush opening. Jesus et al. summarized the methodology for the design of APU air inlets for ERJ145 aircraft. ZHI Wen-jing et al. conducted simulation analysis on the flow field of the inlet system of an APU based on Fluent [8]; DENG Zhi-liang et al. studied the effect of different

T. Zielinska (Ed.): PEAE 2025, LNNS 1679, pp. 64–74, 2026.
https://doi.org/10.1007/978-3-032-08221-3_6

length-width ratios of intake ports on the performance of APU inlet systems under equal area conditions [9]. LI Tian-tian et al. conducted a study on the main influence factors of an APU inlet system based on numerical simulation [11].

APU works mostly when the aircraft is on the ground, and the airport environment affects the working condition of APU [10]. Crosswind is a common atmosphere at airports in northern China. The crosswind affects the inlet of the APU, which further affects its operational performance and may even cause APU surging. This paper mainly studies the impact of crosswind on APU inlet system. The paper first used CATIA to construct a model of an aircraft and its APU inlet system, and then used Star CCM+ to simulate the APU inlet flow field. Finally, the influence of crosswind on the total pressure recovery coefficient and inlet distortion of the APU inlet system was analyzed.

2 Establish the Simulation Model

2.1 Geometry and Meshing

Due to the need for flow field simulation, the original aircraft model needs to be simplified before analysis. This paper used CATIA to extract the surface of the aircraft solid model, and deleted parts that have no impact on the APU inlet flow field, such as the aircraft wing, engine nacelle, etc., as shown in Fig. 2, Similarly, this paper only retained the components that affect the inlet flow field in the APU inlet system, such as inlet door, cooling pipe, and inlet duct, etc., as shown in Fig. 3. The opening of the door was 45°, which was consistent with the opening of the door when APU was working on the ground. The inlet duct is composed of a casing and a silencer which can reduce noise and guide the air. In order to more accurately simulate APU inlet flow field during ground operation, this paper also established a geometry of the APU body's inlet plenum. The APU in this paper contains two compressors located on the front and rear sides of the plenum. The front load compressor provides compressed air to the aircraft, and the rear core compressor provides compressed air to the APU combustion chamber. Therefore, the plenum includes two outlets. At the same time, both outlets of the plenum were stretched to leave sufficient space for the development of small eddies at the mass boundary, which is beneficial for calculation convergence.

The calculation domain selected in this paper was a rectangle, as shown in Fig. 1. Imported the domain, aircraft, and APU geometries into STAR CCM+, and set the target surface mesh sizes of the domain, aircraft and APU to 15 m, 50 mm, and 10 mm, with the minimum surface mesh size to 1.5 m, 10 mm and 1 mm respectively. The mesh models of the domain, aircraft, and APU are shown in Figs. 1, 2, and 3, with a total grid size of 4608630.

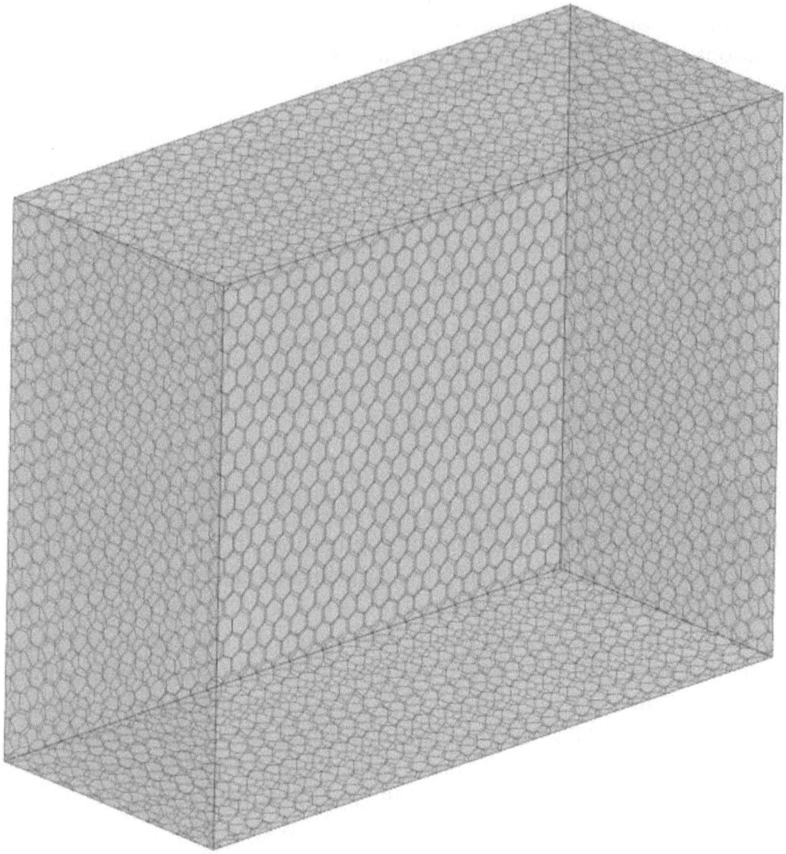

Fig. 1. The mesh model of the domain

Fig. 2. The mesh model of the aircraft

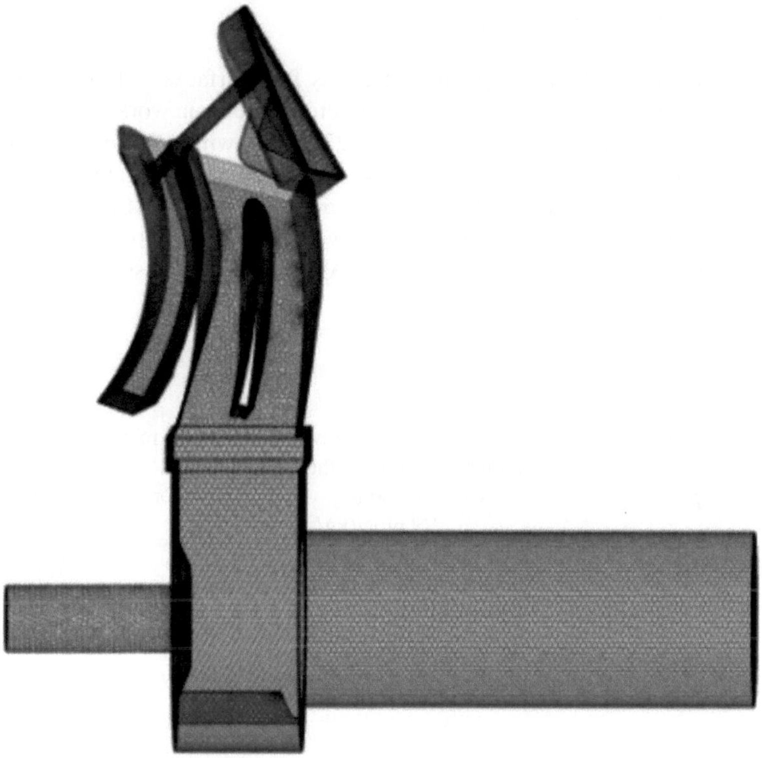

Fig. 3. The mesh model of the APU

2.2 Physical Model

The fluid physical model in this article is shown in Table 1.

Table 1. The fluid physical model

Space	three-dimensional
Time	steady state
Material	gas
Equation of state	ideal gas
Flow	coupled flow
Viscous Regime	turbulence
Reynolds-averaged Turbulence	K-Epsilon turbulence

2.3 Definition of Boundary Conditions

Set the six surfaces of the rectangular domain as free surfaces, set all surfaces wind speed to 18 m/s, and set all surfaces wind direction into four working conditions: 0° head wind, 90° crosswind, 180° tail wind, and 270° crosswind, as shown in Fig. 4;

The calculation condition in this paper is shown in Table 1. The outlet of the cooling pipe and the outlet of the load compressor/core compressor in the APU plenum in Fig. 5 were set as the mass outlet, and the mass flow rates were set to be consistent with Table 2. In order to simulate APU suction in STAR CCM+, the mass flow rates were set to a negative value.

Table 2. The calculation condition

Condition	Ambient pressure /pa	Ambient temperature/°C	Mass flow rate of the cooling pipe/(kg/s)	Mass flow rate of the load compressor /(kg/s)	Mass flow rate of the core compressor /(kg/s)
ground air supply	101325	15	−0.90	−0.91	−1.78

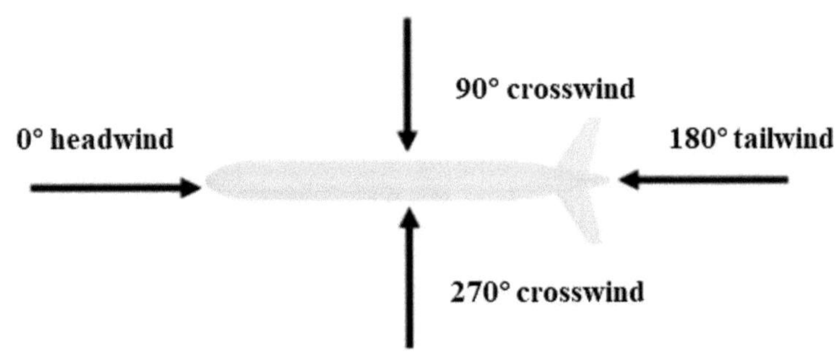

Fig. 4. Different ground wind direction

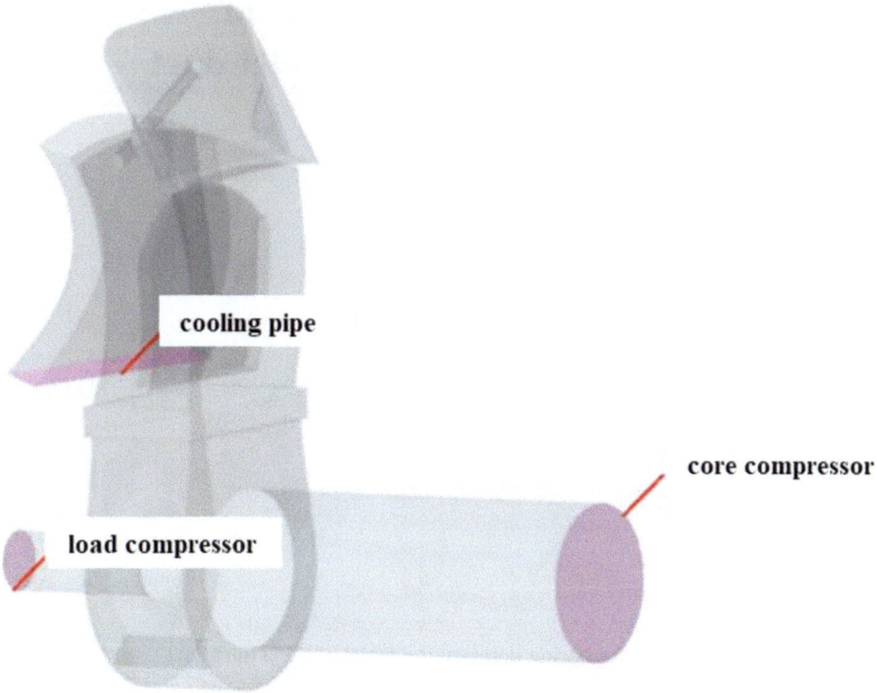

Fig. 5. The outlets of the APU inlet system

2.4 Monitoring Points for the Inlet Distortion

This paper analyzes the influence of crosswind on APU inlet from two dimensions: the total pressure recovery coefficient and the inlet distortion. Therefore, in the calculation process, in addition to monitoring the total pressure at the outlet of APU inlet duct, it is also necessary to monitor the velocity at this outlet. In order to facilitate the monitoring of the velocity, this paper set up 6 columns, 4 rows, and a total of 24 velocity monitoring points at the APU inlet duct outlet, as shown in Fig. 6.

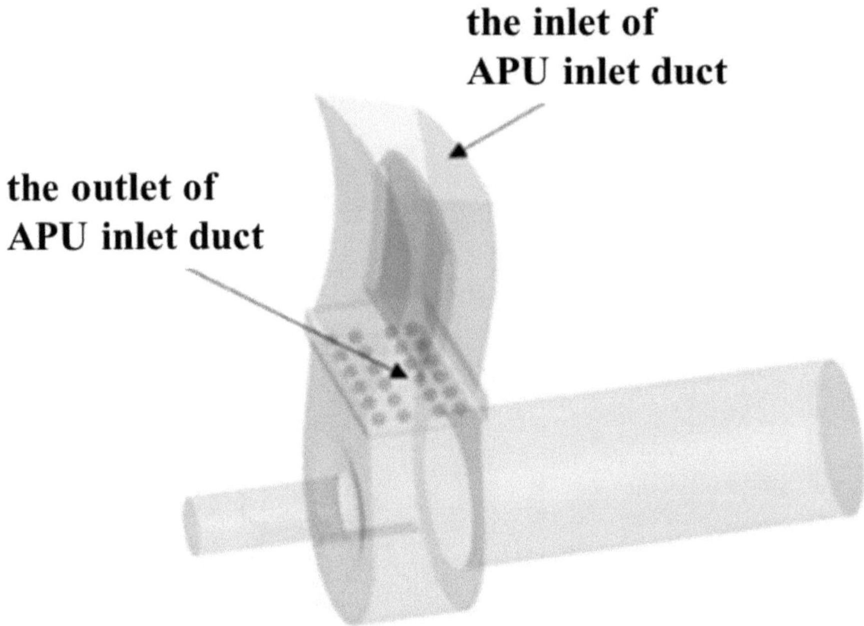

Fig. 6. Monitoring points at the outlet of APU inlet duct

3 Research on the Influence of Ground Wind Direction on the Total Pressure Recovery Coefficient and Distortion of APU Inlet

3.1 The Impact on the Total Pressure Recovery Coefficient

This paper selected the average pressure at the outlet of the APU inlet duct under four wind directions of 0°/90°/180°/270°, and calculated the total pressure recovery coefficient of the APU inlet system under different ground wind directions according to formula 1, as shown in Table 3. Under 180° tail wind, the total pressure recovery coefficient is the lowest, as the inlet of the APU is blocked by the inlet door in this wind direction. The minimum total pressure recovery coefficient of the APU inlet system under different wind directions is 98.9%, and the difference between the maximum and minimum total pressure recovery coefficients is only 0.8%. It indicates that the influence of ground wind direction on the total pressure recovery coefficient of the APU inlet system is very small.

$$C_{inletrecovery} = \frac{P_{Tout}}{P_{ambient}} \tag{1}$$

P_{Tout} is the total pressure at the outlet of the APU inlet duct;

$P_{ambinent}$ is the total environmental pressure, and the total environmental pressure in this paper is 101325pa.

Table 3. The total pressure recovery coefficient under different ground wind directions

Num	Wind direction	total pressure/pa	total pressure recovery coefficient
1	0°headwind	101069	99.7%
2	90°crosswind	100944	99.6%
3	180°tailwind	100201	98.9%
4	270°crosswind	100639	99.3%

3.2 The Impact on the Inlet Distortion

This paper selected the velocities of 24 monitoring points at the outlet of the APU inlet duct under four wind directions of 0°/90°/180°/270°, and calculated the APU inlet distortion under different wind directions according to formula 2, as shown in Table 4. The velocity cloud maps at the APU inlet outlet under different wind directions are shown in Figs. 7, 8, 9 and 10. Under 0° headwind, the APU inlet is the most uniform with small clockwise distortion; Under 180° tail wind, the clockwise APU inlet distortion increases. Crosswind not only significantly increases the distribution unevenness of APU inlet, but also changes the direction of APU inlet distortion, changing clockwise distortion to counterclockwise distortion. The distortion under 90° crosswind is significantly greater than the distortion under 270° crosswind, because the APU intake is placed on the right side of the aircraft, and the 90° crosswind intake is not obstructed by the aircraft.

$$DI = \frac{\sum_{i=1}^{24} v_i^2 r_i}{\sum_{i=1}^{24} v_i} \tag{2}$$

v_i is the velocity magnitude of each measuring point at the outlet of APU inlet duct;

r_i is the distance between each measuring point and the centerline of the outlet, where the distance from measuring points 1 to 12 is set as a positive value, and the distance from measuring points 13 to 24 is set as a negative value;

DI represents the distortion of the APU inlet distortion, with positive values indicating clockwise rotation of the airflow in the APU plenum and negative values indicating counterclockwise rotation of the airflow.

Table 4. The APU inlet distortion under different wind directions

Num	Wind direction	APU inlet distortion/(m^2/sec)
1	0°headwind	0.57
2	90°crosswind	−1.65
3	180°tailwind	0.76
4	270°crosswind	−1.14

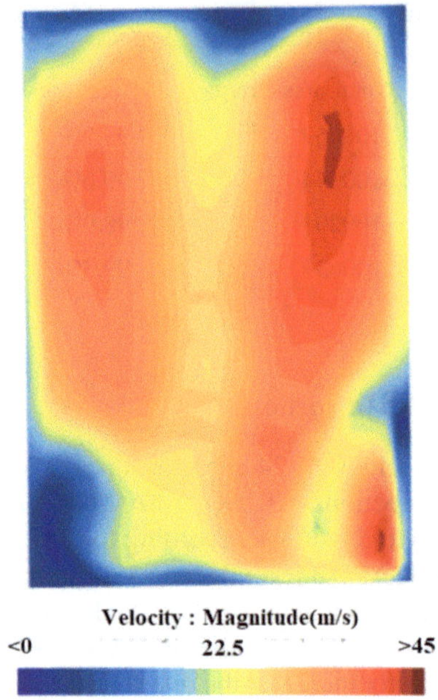

Fig. 7. The velocity cloud maps at the APU inlet outlet under 0°headwind

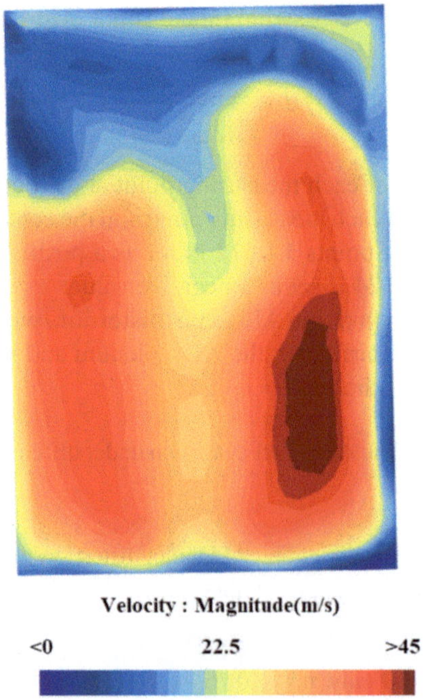

Fig. 8. The velocity cloud maps at the APU inlet outlet under 90°crosswind

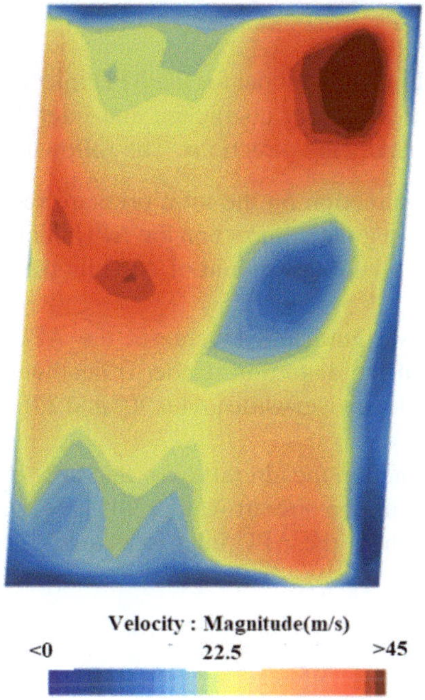

Fig. 9. The velocity cloud maps at the APU inlet outlet under 180° tailwind

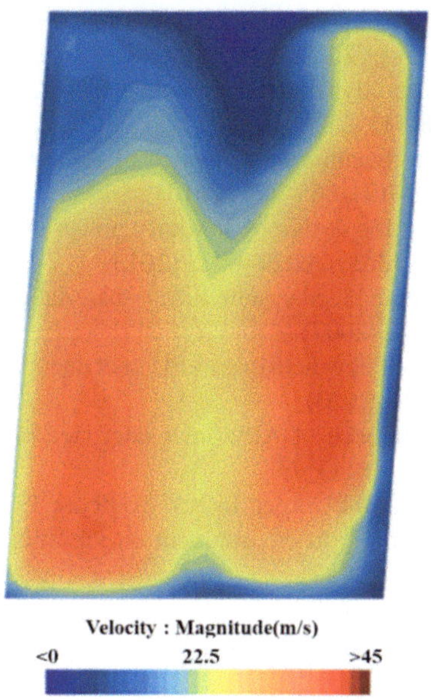

Fig. 10. The velocity cloud maps at the APU inlet outlet under 270° crosswind

4 Conclusion

This paper established a simulation model for the APU inlet flow field of an aircraft, and compared the total pressure recovery coefficient and inlet distortion of the APU inlet system under different ground wind directions. The following conclusions are drawn:

1) The influence of wind direction on the total pressure recovery coefficient of APU intake is small, but it has a greater impact on the APU inlet distortion. It indicates that crosswind doesn't affect the total amount of APU intake, but affects the uniformity of APU intake.
2) The crosswind not only significantly increases the APU inlet distortion, but also changes the direction of the distortion. Under 0°headwinds and 180 tailwinds, the APU inlet distortion is clockwise, while under 90° and 270°crosswinds, the distortion is counterclockwise.
3) When the APU inlet is not on the aircraft axis, the APU inlet distortion is related to the position of the inlet on the aircraft. Under 90° crosswind (the crosswind comes from the aircraft side where the APU inlet is located), the APU inlet distortion is maximum.

References

1. Zhao, Z.: Study on the APU control system of civil aircraft. Appl. Energy Technol. (5), 17–19 (2017)
2. Huang, G.: Comparison of technical characteristics of APU and micro turbine engines for large aircraft. J. Aerosp. Power **23**(2), 383–388 (2008)
3. Wang, M.: Experiment on aerodynamic performance of APU air inlet system with perforated plate. J. Aerosp. Power **28**(11), 2526–2535 (2013)
4. Wang, H.: A study on aerodynamic performance of APU air inlet system with perforated plates. J. Propul. Technol. **34**(7), 865–872 (2013)
5. Gao, P.: A methodology for the design of auxiliary power unit air inlet system for transport aircraft. Sci. Technol. Eng. **13**(20), 6020–6023 (2013)
6. Adib, A.M.L.: Aircraft engine bleed system tubes: material and failure mode analysis. Eng. Fail. Anal. **14**(8), 1605–1617 (2007)
7. Zhi, W.: Numerical analysis of flow field in an APU inlet duct based on fluent. Chin. Hydraulics Pneumatics **47**(10), 119–126 (2023)
8. Deng, Z.: Study on the influence of APU inlet duct size ratio on inlet performance. Sci. Technol. Inf. (22), 447–449 (2011)
9. Li, T.: Optimization design of APU inlet. Intern. Combust. Engine Parts **6**(36), 1–3 (2022)
10. Lv, J.: Influence analysis on airport environment on vibration of civil aircraft APU. Mach. Build. Autom. (04), 224–228 (2022)

Modern Mechanical System and Reliability Evaluation

Safety Analyses of Cylinder Ejection Process in Different Conditions

Yan Huo[1], Jianhong Sun[1,2](✉), Daren Zheng[1], Weidong Zhao[1], Zhi Sun[2], and Pei Li[1]

[1] Key Laboratory of Aircraft Environment Control and Life Support, MIIT, Nanjing University of Aeronautics and Astronautics, Nanjing, China
{nuaahy,jhsun,zhaowd,lp2020}@nuaa.edu.cn

[2] Key Laboratory of Civil Aviation Emergency Science and Technology, CAAC, Nanjing University of Aeronautics and Astronautics, Nanjing, China
z.sunny@nuaa.edu.cn

Abstract. The off-plane separation technology of rigid body projectiles is one of the key technologies for the development of a new generation of aircraft. The off-plane safety analysis of rigid body projectiles in different ejection conditions remains to be investigated. At present, the research on the effect of the angle of attack and the ejection angle of the cylindrical projectile on the off-plane safety of cylindrical projectile is still insufficient and requires further research. The off-plane safety performance of the cylindrical projectile is investigated in this paper. The ejection separation trajectory and off-plane safety are simulated by solving the unsteady Navier-Stokes equations and the k-ω Shear Stress Transport turbulence model coupled with the six-degree-of-freedom motion equation. The effect of the ejection angle and the aircraft angle of attack on the off-plane safety of the cylindrical projectile is evaluated by demonstrating distributional properties. The results obtained show that the cylindrical projectile rotate after leaving the aircraft, and the trajectory of the cylindrical projectile during separation is parabolic. The separation trajectory tends to be a straight line as the angle of attack of the aircraft increases. The off-plane safety can be improved by reducing the angle of attack of the aircraft. Furthermore, the off-plane safety of cylindrical projectile improves as the ejection angle increases.

Keywords: off-plane safety · the aircraft angle of attack · the ejection angle · cylindrical projectile

1 Introduction

In aviation and aerospace technology, there is a common issue with relative motion between multiple objects. The ejection separation motion as a multi-body separation issue, due to the instability of the flow field during separation, multi-body separation motion has a great impact on the safety of the aircraft, so the study of the effect of the separation parameters on the off-plane safety of cylindrical projectile is a necessary topic. Multi-body separation can be observed in launch vehicles, air-to-air missiles,

T. Zielinska (Ed.): PEAE 2025, LNNS 1679, pp. 77–89, 2026.
https://doi.org/10.1007/978-3-032-08221-3_7

paratrooper jumps, heavy airdrops, and ejection seats. Due to the strong, nonlinear and unsteady aerodynamic forces interfere with the multibody motion issue, an important methodology to solve the multibody separation issue is the coupling between the flow field and the motion [1–3]. The flight of the cylindrical projectile may be abnormal due to the complexity of the flow field, which could damage the aircraft's safety. The separation trajectory of the cylindrical projectile must be safe and will not pose any threat to the aircraft, and the short-term separation motion will not affect the separation ability to complete the expected mission [4]. The off-plane safety of cylindrical projectile is primarily concerned with the effect of cylindrical projectile on flight safety when airflow disturbances occur after ejection separation [5]. Based on the coupling method of computational fluid dynamics (CFD), rigid body dynamics (RBD) and computational structural dynamics (CSD), Tian found that Mach number and flight height are two important flight parameters affecting the separation of missile and inner cabin [6]. Through the coupling method of unsteady N-S equation and rigid body six-degree-of-freedom equation, Zhou found that the flight at positive angle of attack has little effect on missile separation, and the flight at negative angle of attack will lead to the yaw of missile trajectory [7]. Johnson found that the separation trajectory of the missile will change with the launch time under the effect of the unsteady flow field through the RANS method and wind tunnel test [8, 9]. Westmoreland's research shows that applying an appropriate ejector effect to the released missile can improve its separation trajectory [10]. Olejnik analyzed the behavior of free-falling bodies under aerodynamic interference using 6-DOF motion and CFD in combination with simulation. The objective is to demonstrate that this type of research is highly significant in improving the safety level of flight operations and preventing hazardous situations in the air [11]. The modal method was used by Hua to combine the static aeroelastic equation with the 6-DOF dynamic equation and examine the impact of elastic deformation on the trajectory of mid-air separation. The study concluded that attitude angle oscillation can be reduced by elastic deformation [12]. Zhang Yantai employed the Arbitrary Lagrange-Euler (ALE) method to study the parachute deployment process during the ejection of the civil aviation separated emergency data recorder. By examining the impact of incoming velocity and angle on the parachute canopy's inflation, the findings indicate that the parachute's ejection velocity can be safely and moderately increased to enhance safety during the deployment process [13]. Feng Chuanqi used the Arbitrary Lagrange Euler (ALE) finite element method to numerically simulate the inflation process of a certain type of heavy airdrop parachute. Based on the dynamic equations and stability analysis theory of multi-body systems, further in-depth research is conducted on the trajectory and stochastic disturbances of the opening process of the finite mass parachute [14].

At present, the research on the effect of the angle of attack and the ejection angle of the cylindrical projectile on the off-plane safety of cylindrical projectile is still insufficient and requires further research. In order to study the effect of separation parameters on off-plane safety, this paper adopts a coupling simulation method of CFD and six-degree-of-freedom motion (6-DOF motion) equations to simulate the ejection separation process and analyze the off-plane safety and the separation trajectory under different operating conditions. The separation process of cylindrical projectile ejected from an airplane can be simulated using a mixture of unsteady N-S equations, rigid body equations of motion, and k-ω SST turbulence model. To investigate the safety of ejection separation, the separation parameters are analyzed to examine separation trajectories, velocities, and accelerations of the cylindrical projectile.

2 Numerical Model

Reynold's number represents the ratio of inertial force to viscous force and also characterizes the flow as laminar, transient, or turbulent [15]. When the Reynold number is below the lower critical value, the flow is laminar; when it exceeds the upper critical value, it becomes turbulent. When the Reynold number between these two limits, the flow is transient. The formula for Reynold's number can be expressed as:

$$R = \frac{\nu L}{\mu_{\text{air}}} \tag{1}$$

where R is the value of Reynold's number, ν is the airplane's speed, L is the length of the airplane, and μ_{air} is the viscosity of air.

The off-plane safety of cylindrical projectile in the case of civil aircraft cruising is analyzed in this paper. Since the civil aircraft's cruising Mach number is larger than 0.3, air compressibility cannot be disregarded. The compressible Navier-Stoke (N-S) equation is used to simulate the cylindrical projectile's separation process [7]. The following is an expression for the governing equation:

$$\frac{\partial \rho}{\partial t} + \frac{\partial U_i}{\partial x_i} = 0 \tag{2}$$

$$\rho \frac{\partial U_i}{\partial t} + \rho \frac{\partial U_i U_j}{\partial x_j} = -\frac{\partial p}{\partial x_i} + \frac{\partial}{\partial x_{ij}} \left[(\mu + \mu_T) \left(\frac{\partial U_i}{\partial x_j} + \frac{\partial U_j}{\partial x_i} \right) \right] \tag{3}$$

where, U_i represents the velocity vector in Cartesian coordinate and x_i is Cartesian coordinate. t is the time, ρ is the fluid density and p is the dynamic pressure. In addition, μ is the effective viscosity coefficient of dynamic viscosity and μ_T represents the eddy viscosity.

In this paper, the computational fluid dynamics is used for numerical simulation. The k-ω Shear Stress Transport (SST) turbulence model is used to solve the viscosity issue. Concretely, the transport equations of k and ω are written as:

$$\rho \frac{\partial k}{\partial t} + \rho \frac{\partial U_j k}{\partial x_j} - \frac{\partial}{\partial x_j} \left[(\mu + \sigma_k \mu_T) \frac{\partial k}{\partial x_j} \right] = \tau_{ij} \frac{\partial U_i}{\partial x_i} - \beta^* \rho k \omega \tag{4}$$

$$\rho \frac{\partial \omega}{\partial t} + \rho \frac{\partial U_j \omega}{\partial x_j} - \frac{\partial}{\partial x_j} \left[(\mu + \sigma_\omega \mu_T) \frac{\partial \omega}{\partial x_j} \right] = \frac{\gamma}{\nu_T} \tau_{ij} \frac{\partial U_i}{\partial x_i} - \beta \omega^2 + 2\rho (1 - F_1) \frac{\sigma_{\omega 2}}{\omega} \frac{\partial k}{\partial x_j} \frac{\partial \omega}{\partial x_j} \tag{5}$$

where, k is turbulent kinetic energy and ω is the specific dissipation of turbulent kinetic energy. The transport equations of k and ω are written as:

$$k = \frac{1}{2} u_i' u_j' \tag{6}$$

$$\omega = \frac{\varepsilon}{k\beta^*} \tag{7}$$

where ε is the turbulent kinetic energy dissipation rate, which is defined as:

$$\varepsilon = \nu_T \frac{\partial u_i'}{\partial x_k} \frac{\partial u_i'}{\partial x_k} \tag{8}$$

where τ_{ij} is Reynold stress, which is defined as:

$$\tau_{ij} = \mu_T \left(\frac{\partial u_i}{\partial x_j} + \frac{\partial u_j}{\partial x_i}\right) - \frac{2}{3}\rho\delta_{ij} \tag{9}$$

where δ_{ij} is the Kronecker function, $\nu_T = \mu_T/\rho$ is dynamic eddy viscosity and F_1 is a mixed function. σ_k, β^*, σ_ω, γ, β and $\sigma_{\omega 2}$ are the model coefficients.

3 Simulation Model and Separation Position

The civil aircraft is used as an aircraft model to analyze the influence of separation parameters on the safety of departure, as depicted in Fig. 1. In order to ensure the safety of the separation as much as possible, the effect of the aircraft on the off-plane safety should be reduced as much as possible, so the airborne ejection system can be put behind the spherical frame [16]. The location of the cylindrical projectile is shown in Fig. 2. The projectile of this simulation example is a cylinder with a diameter of 140 mm and a height of 150 mm.

Fig. 1. The civil aircraft

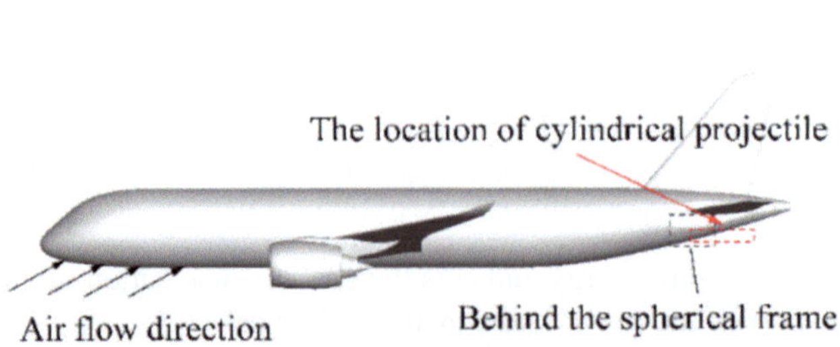

Fig. 2. The cylindrical projectile location.

4 Computation Domain, Boundary Conditions and Mesh

Overset mesh was applied to CFD for its accuracy, efficiency, and high-adaptability to the wide variations of trim and heave [17]. Figure 3 depicts the computational domain, and Fig. 4 displays the boundary conditions and the overset region, where the dimensions are referred to as the length of the fuselage, L. The inflow velocity for this study is 120 m/s, the separator's ejector velocity is 10 m/s, and the separator weighs 6 kg. By altering the ejection angle and attack angle, the effect of different separation parameters on off-plane safety is evaluated in this paper.

The flow computation and the trajectory simulation are decoupled over a small time steps. At each time step, the flow computation is carried out until convergence of the flow field to reach the steady-state solution for specified missile position from the aircraft. Then, the 6-DOF solver is applied to obtain the new location of the missile from the aircraft. The selection of time step is very important because the larger time steps lead to error that accumulates over time, whereas the smaller time steps take considerably longer simulation time for various trajectory-simulation cases.

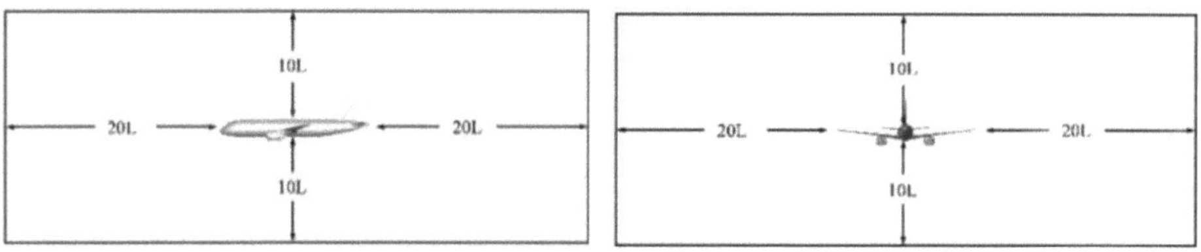

Fig. 3. Computational domain dimensions

The primary component of the cutting body grid is the orthogonal right-angle hexahedron unit, which is employed for both the cutting encryption and the generation of a boundary layer mesh. The latter is achieved through the use of an octree or composite fork tree, which offers high encryption efficiency and overall good performance. The cutting body mesh can be fitted with different forms of unstructured units, thereby demonstrating strong adaptability and the capacity to generate a boundary layer mesh [18].

A changing overset region and a static backdrop region were merged in the computational domain. The entire computational domain was surrounded by the backdrop, and the overset region moved alongside the segregator as a rigid body to maintain grid quality close to the segregator. The ejection separation process involves translation and rolling motion in all directions, so the six-degree-of-freedom function of DFBI in the software is enabled. At the same time, the overlapping grid method is adopted. At the overset boundary, the acceptor cells are used to couple the solutions on the two regions, and divide active and inactive cells in the background region. The discretized governing equations are solved within active cells which update with the moving overset region. To improve the accuracy of interpolation, the overlapping interface in the background should be refined to match the size and the distribution of the grid at the interface of overset region.

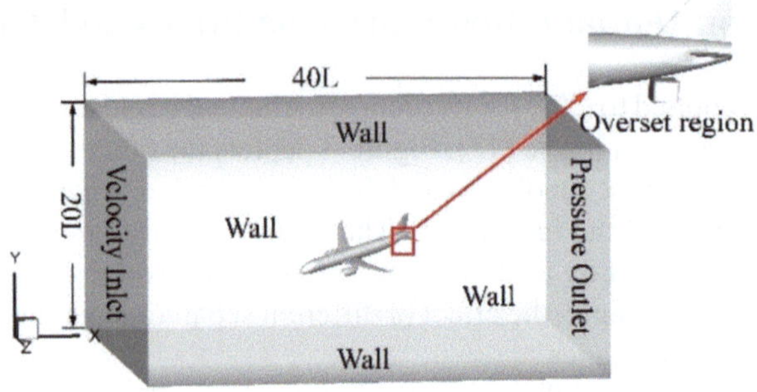

Fig. 4. Boundary conditions

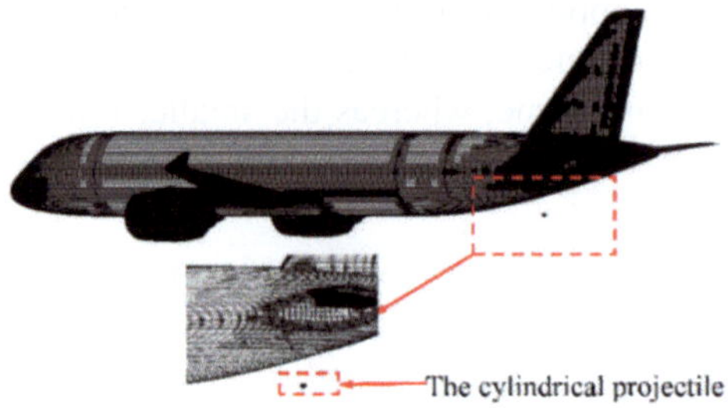

Fig. 5. The civil aircraft surface grid and the location of cylindrical projectile.

The airplane surface mesh for the civil aircraft encrypting the position of the wing, tail, vertical tail, and engine of the civil aircraft is done. Figure 5 displays the location of the cylindrical projectile and aircraft surface mesh for the civil aircraft model. The flow field domain grid is refined in the vicinity of the aircraft, and Fig. 6 (a) displays the flow field domain grid as a whole. Figure 6(b) displays the fuselage symmetry plane's starting grid. Figure 6(c) depicts the fuselage encryption area's mesh, which has 75 million grids overall.

(a). The whole grid of the domain

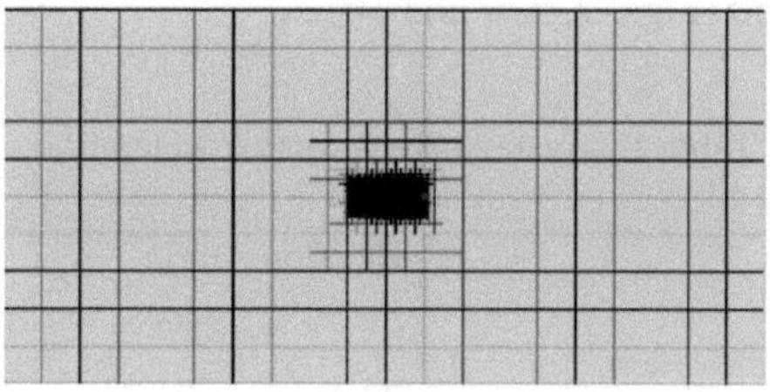

(b). The initial grid of the fuselage symmetry plane

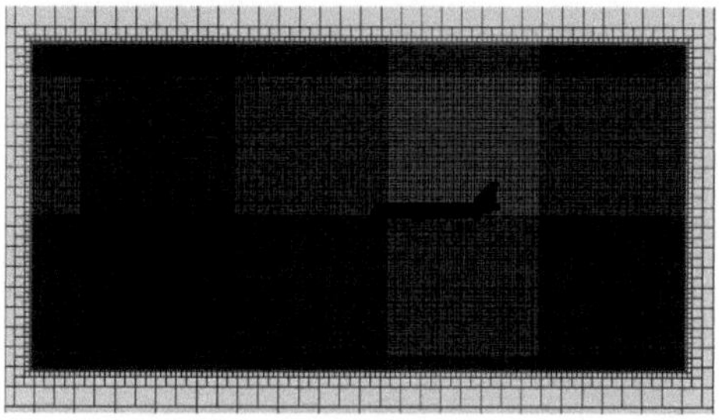

(c). The grid of the fuselage encryption area

Fig. 6. Flow field grid

5 Result and Analysis

The purpose of this paper is to analyze and discuss the separation characteristics of rigid objects by examining the effect of separation trajectory under different separation parameters. In order to more clearly show the effect of separation parameters on the separation trajectory and off-plane safety of the cylindrical projectile, the effect of separation angle and the angle of attack of aircraft are studied.

5.1 Effect of Ejection Angel

The effect of four ejection angles (angle between ejection direction and x-axis) on the safety of the cylindrical projectile were analyzed in this study. The ejection angles were 0°, 45°, 76.94° (vertical belly) and 90° respectively. The trajectory of cylindrical projectile takes the shape of a parabola. As the ejection angle increases, the separation moves away from the aircraft at a faster pace. The trajectory of the cylindrical projectile shows that the parabolic trajectory opening is upward when the ejection angle is more than or equal to 45° and downward when the ejection angle is 0°. Because the cylindrical projectile's acceleration is always higher than zero, the cylindrical projectile's horizontal velocity increases with time after the cylindrical projectile's ejection from the aircraft, and the acceleration curve fluctuates around the average figure. The Table 1 shows that although the average acceleration increases as the ejection angle increases, and the average acceleration decreases as the ejection angle reaches 90°. The acceleration plot shows that when the ejection angle rises, the angular velocity curve's peak figure progressively falls behind (Figs. 7, 8, 9 and 10).

Table 1. Average Horizontal Acceleration

	Ejection Angle			
	0°	*45°*	*76.94°*	*90°*
Average Horizontal Acceleration (m/s^2)	22.25	23.74	24.81	19.65

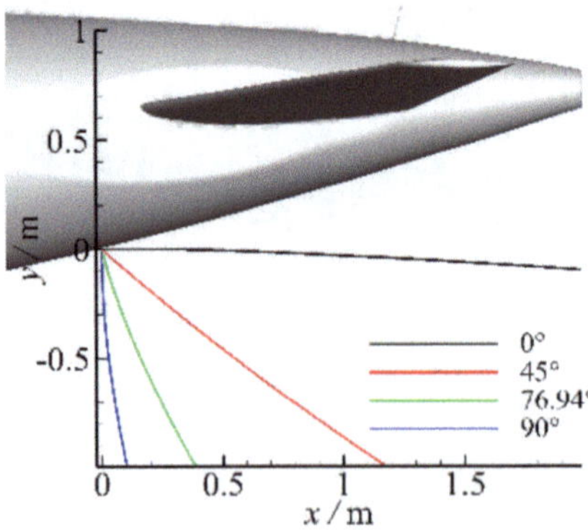

Fig. 7. Separation trajectory of cylindrical projectile.

The size of the initial force area of the cylindrical projectile change with the change of the ejection angle, so the horizontal force of the cylindrical projectile are different accordingly, so the initial accelerations of the cylindrical projectile are also different. Therefore, no similar characteristic law between the acceleration and angular velocity curves of the cylindrical projectile.

In order to improve the off-plane safety of the cylindrical projectile, the cylindrical projectile should be ejected with a large ejection angle. When the large ejection angle is adopted, the separation trajectory of the cylindrical projectile is far away from the

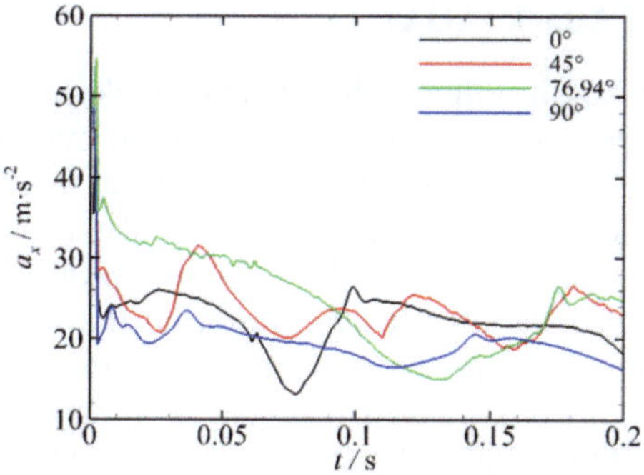

Fig. 8. The horizontal acceleration of cylindrical projectile

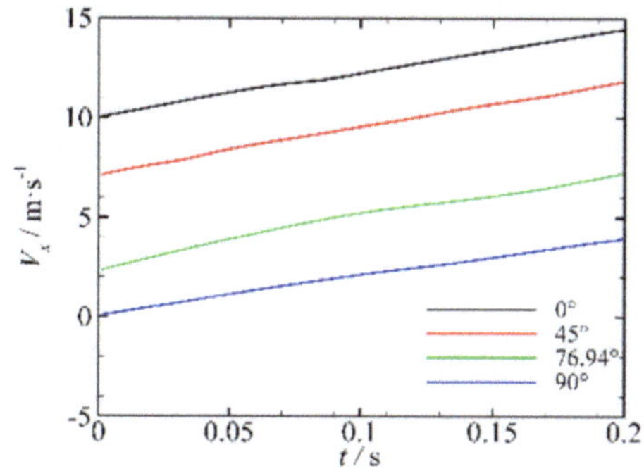

Fig. 9. The horizontal velocity of cylindrical projectile.

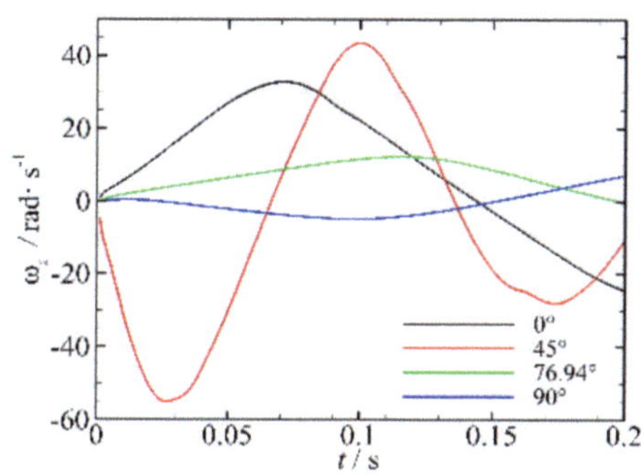

Fig. 10. The angular velocity of cylindrical projectile

aircraft, and the horizontal acceleration and angular velocity of the cylindrical projectile are relatively stable.

5.2 Effect of Angel of Attack (AOA)

This paper mainly analyzes the effect of four aircraft angles of attack on the separation safety. The aircraft angles of attack are 0°, 10°, 20° and 30° respectively. As the angle of attack decreases, the separation trajectory gradually moves away from the aircraft. The cylindrical projectile's separation trajectory is shaped like a parabola. As the aircraft's angle of attack increases, the trajectory of the cylindrical projectile tends to be a straight line. As the angle of attack of the aircraft increases during the first 0.12 s of the separation, the cylindrical projectile also experience an increase in the horizontal velocity. From the horizontal velocity of the separated material, at a time of 0.075 s, the slope of the horizontal velocity of the separated material decreases. After 0.075 s, the cylindrical projectile's horizontal velocity drops very significantly when the aircraft's angle of attack is 20°and 30°. The cylindrical projectile's velocity is less than the cylindrical projectile's while the aircraft is attacking at an angle of 0° and 10° after 0.12 s (Figs. 11, 12, 13 and 14).

Table 2. Average Horizontal Acceleration

	Angle of Attack			
	0°	*10°*	*20°*	*30°*
Average Horizontal Acceleration (m/s^2)	22.25	22.14	22.01	20.78

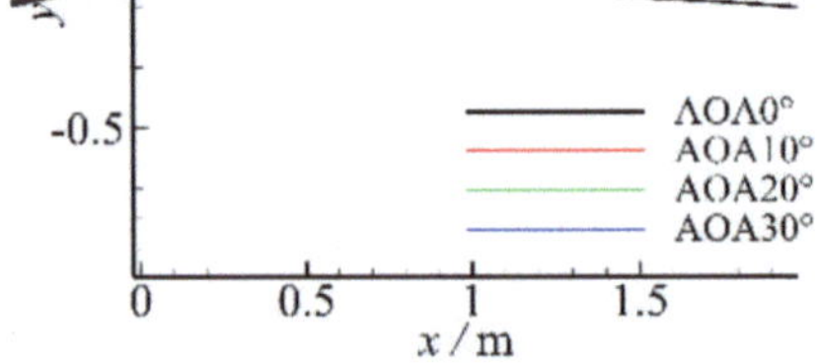

Fig. 11. Separation trajectory of cylindrical projectile

The time taken for the acceleration to reach its minimum and maximum decreases as the aircraft's angle of attack increases. When the time is greater than 0.1 s, the horizontal acceleration of the separation decreases with the angle of attack of the aircraft increases. According to Table 2, the average horizontal acceleration of the separation decreases as the angle of attack of the aircraft increases. Since the figure of the horizontal acceleration

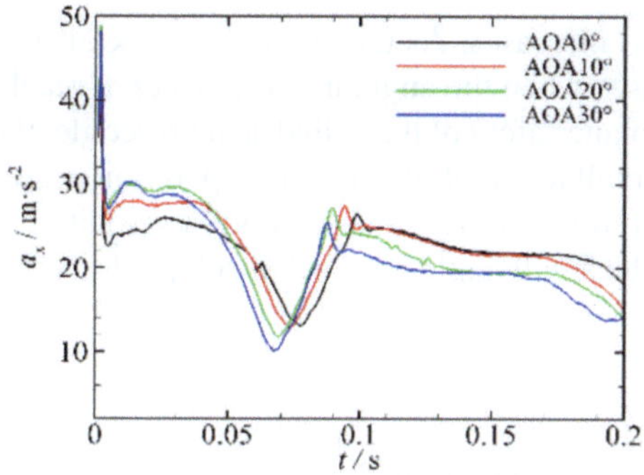

Fig. 12. The horizontal acceleration of cylindrical projectile

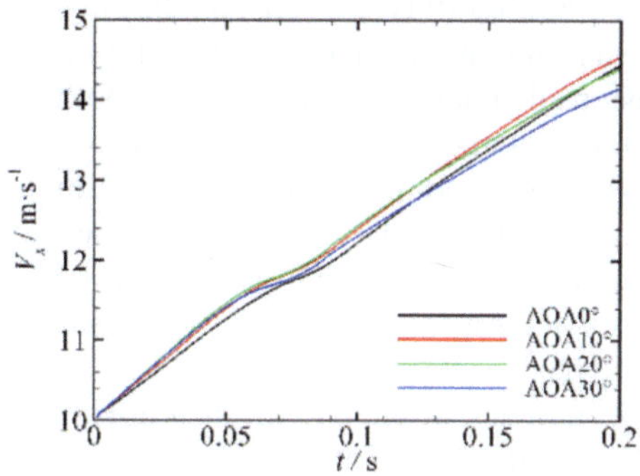

Fig. 13. The horizontal velocity of cylindrical projectile.

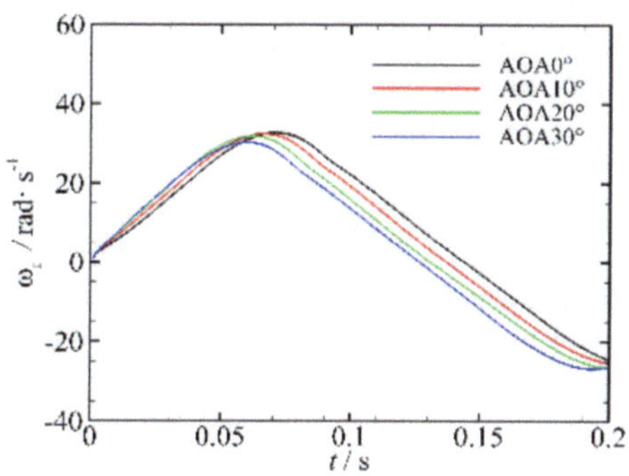

Fig. 14. The angular velocity of cylindrical projectile.

of the cylindrical projectile is always higher than zero, the horizontal velocity of the separation increases with time. Along with the angle of attack of the aircraft increases, the angular velocity of the separation can reach the peak and trough of the peak faster, and the peak and trough are smaller. Before the angular velocity curve reaches the peak, the angular velocity increases with the angle of attack of the aircraft increases, and the angular velocity also increases. As the angle of attack of the aircraft increases, the angular velocity decreases when the angular velocity curve reaches its peak.

To improve the off-plane safety of the cylindrical projectile, the cylindrical projectile should be ejected at a small angle of attack. The separation trajectory of the cylindrical projectile is far away from the aircraft, the separation velocity is stable and the angular velocity is large when the small angle of attack is adopted.

6 Conclusion

In this study multi-body motion is studied by solving the unsteady Navier-Stokes equation in the flow field and the six-degree-of-freedom motion equation of the rigid body based on the overlapping grid. The simulation of the separation of the cylindrical projectile in the tail section of the aircraft is realized, and the effect of the ejection angle and the angle of attack of the aircraft on the separation trajectory of the cylindrical projectile are studied. The conclusions that have been drawn are as follows:

The average horizontal acceleration of the cylindrical projectile decreases as the angle of attack of the aircraft increases and the average acceleration increases as the ejection angle increases, and the average acceleration decreases as the ejection angle reaches 90°. As the ejection angle increases, the separation moves away from the aircraft at a faster pace. As the angle of attack of the aircraft increases, the angular velocity decreases when the angular velocity curve reaches its peak.

Along with the angle of attack of the aircraft increases, the angular velocity of the separation can reach the peak and trough of the peak faster, and the peak and trough are smaller. As the angle of attack of the aircraft increases, the cylindrical projectile tend to have a straight separation trajectory instead of the parabolic one. The safety of leaving the aircraft can be improved by reducing the angle of attack of the aircraft. With the ejection angle increases, so does the off-plane safety of cylindrical projectile. Safety and reliability during the separation process can be ensured by using a large ejection angle even with a small aircraft angle of attack.

Acknowledgment. The authors gratefully acknowledge the support of Shanghai Central Government Leading Local Science and Technology Development Funds (Grant No. YDZX20233100004008), the Fundamental Research Funds for the Central Universities: Key Laboratory of Civil Aviation Emergency Science and Technology CAAC (Grant No. XCA2402202).

References

1. Toro, E.F., Chakraborty, A.: The development of a Riemann solver for the steady supersonic Euler equations. Aeronaut. J. **98**(979), 325–339 (1994)
2. Zhu, S., Chen, Z., Zhang, H., et al.: Investigations on the influence of control devices to the separation characteristics of a missile from the internal weapons bay. J. Mech. Sci. Technol. **32**, 2047–2057 (2018)
3. Taylor, N.: Separated Flow: Some Challenges and Research Priorities for Missile Aerodynamics. Separated Flow: Prediction, Measurement and Assessment for Air and Sea Vehicles; NATO Science and Technology Organization: Neuilly-sur-Seine, France (2019)

4. Anandhanarayanan, K., Raj, A., Shah, V., et al.: Separation dynamics of air-to-air missile and validation with flight data. Defence Sci. J. **68**(1) (2018)
5. Zheng, S.: Study on safety of missile separated from craft in launch. J. Sichuan Ordnance **2015**(6), 21–24 (2015)
6. Tian, S., Li, R., Xu, K.: Investigation of aeroelasticity effect on missile separation from the internal bay. Int. J. Aerosp. Eng. **2023**(1), 9875622 (2023)
7. Wei, Z.; Research on the Influence of Flight Attitude on Unilateral Missile Separation from Embedded Bay. September 4–9. Stockholm, Sweden (2022)
8. Johnson, R., Stanek, M., Grove, J.: Store separation trajectory deviations due to unsteady weapons bay aerodynamics. In: 46th AIAA Aerospace Sciences Meeting and Exhibit, p. 188 (2008)
9. Davis, M., Yagle, P., Smith, B., et al.: Store trajectory response to unsteady weapons bay flowfields. In: 47th AIAA Aerospace Sciences Meeting including The New Horizons Forum and Aerospace Exposition, p. 547 (2009)
10. Westmoreland, W.: Trajectory variation due to an unsteady flow-field. In: 47th AIAA Aerospace Sciences Meeting including The New Horizons Forum and Aerospace Exposition, p. 550 (2009)
11. Olejnik, A., Dziubiński, A., Kiszkowiak, Ł.: Separation safety analysis using CFD simulation and remeshing. Aerosp. Sci. Technol. (2020)
12. Hua, R., Zhao, C., Ye, Z., et al.: Effect of elastic deformation on the trajectory of aerial separation. Aerosp. Sci. Technol. **45**, 128–139 (2015)
13. Zhang, Y., Sun, J., Hou, B., et al.: The inflation process and safety analyses of a parachute ejected from civil aircrafts. Acta Aerodynamica Sinica **40**(2), 79–87 (2022)
14. Feng, C., Liu, H., Sun, J., et al.: The study on parachute inflation process of heavy-load airdrop based on ALE. In: Chinese Society of Mechanics, Zhejiang University. Chinese Mechanics Conference Proceedings, p. 7 (2019)
15. Lutfi, M., Prabowo, R., Muslimy, E., et al.: Leisure boat concept design: study on the influence of hull form and dimension to increase hydrodynamic performance. Int. J. Mech. Eng. Robot. Res. **13**(1) (2024)
16. Zhang, Y., Sun, J., Hou, B., et al.: Design and test of the ejection emergency flight data recording and tracking system. J. Beijing Univ. Aeronaut. Astronaut. **47**(11), 2322–2330 (2021). (in Chinese)
17. Adanta, D., Fattah, I.M.R., Muhammad, N.M.: Comparison of standard k-epsilon and SST k-omega turbulence model for breastshot waterwheel simulation. J. Mech. Sci. Eng. (2020)
18. Gong, T., Hu, S., Li, Y.: The investigation of polyhedral mesh and trim mesh on aerodynamic calculation accuracy of hypersonic aerospace plane. J. Nanchang Hangkong Univ. Nat. Sci. (2023)

Aircraft Canopy Jettisoning Shock Waves for Auditory Organ During Ejection Rescue

Jun Li(✉), Wujun Xie, Hui Li, and Liyi Fang

Key Laboratory of Science and Technology on Life-Support Technology, AVIC Aerospace Life-Support Industries LTD., Xiangyang, Hubei, China
1273987282@qq.com

Abstract. The prerequisite for ejection rescue is the clearance of the ejection path, with cockpit canopy jettisoning being a crucial method for path clearance. Research on the shock waves produced during jettisoning is essential for protecting pilot auditory organs. Ground-based experiments on canopy jettisoning were designed, obtaining pressure data from dummy ears using shock wave testing equipment. Auditory organ injury safety limits were calculated based on sound pressure pulse width and national standards, employing the VMD and Hilbert transform (VMD-HT) to analyze time-frequency characteristics of canopy jettisoning shock waves. The results indicate that shock wave energy is concentrated in low frequency region under 5 kHz, lasting no more than 5 ms. Current pilot helmets meet the protection requirements against canopy jettisoning shock waves, yet improvements based on the shock waves' time-frequency characteristics are recommended.

Keywords: Canopy Jettisoning · Shock Waves · Auditory Organs · Safety Limits · VMD-HT

1 Introduction

As aircraft speeds increase, the challenges and risks faced by pilots escalate accordingly. Ejection rescue technology, essential for safeguarding pilot lives, holds a pivotal position worldwide [1]. During flight, when an aircraft malfunctions and a pilot must parachute, the immediate priority is to swiftly jettison the cockpit canopy. Only after clearing the ejection path from the seat, it is safe to initiate ejection. Current research on cockpit canopy jettisoning focuses on clearing ejection paths efficiently [2]. The power source for cockpit canopy jettison systems is typically separate from other aircraft power sources, using independent high-pressure cold air, high-energy jettisoning charges, or cockpit canopy rockets [3]. During jettisoning, shock waves pose a safety risk to human auditory organs, which are susceptible to injury [4–7]. Therefore, analyzing and testing the shock waves produced during canopy jettisoning is critical for developing pilot auditory protection equipment.

While specific literature on canopy jettisoning shock waves are not found, recent studies on shock wave analysis methods and protective measures during ejection rescue

T. Zielinska (Ed.): PEAE 2025, LNNS 1679, pp. 90–97, 2026.
https://doi.org/10.1007/978-3-032-08221-3_8

processes [4–7] are highly relevant. Guan H discussed principles of canopy-jettisoning and canopy-penetration clearance methods, analyzed damage limit calculation methods for auditory and non-auditory organs in China and the USA, and argued for several important research directions for future pilot exposure to explosive shock waves [4]. Xie W studied the impact of pulse noise and shock waves generated by parachute ejection on pilot auditory organs, demonstrating through wavelet analysis the protective effect of helmets against noise in different frequency bands [5]. [6] illustrated that the VMD algorithm is superior to wavelet analysis and EMD algorithms in parachute gun muzzle noise decomposition, avoiding mode overlap, without requiring empirical analysis of basis functions; VMD-HT time-frequency analysis results indicated significant energy concentration of the 3 ms duration seat parachute gun shock wave signal in a low-frequency bandwidth for 0 to 1 kHz. [7] calculated the safe limit of explosive shock waves endured by pilots during explosive cutting canopy for ejection path clearance, and laboratory experiments with various protective materials demonstrated that rubber materials provide better protection for pilots.

GJB349.28 stipulates the measurement technical requirements and methods of muzzle shock wave overpressure (the peak overpressure between 6.9 kPa and 100 kPa) during artillery firing, which provides unified technical measures for experimental research, design and finalisation of pyrotechnic products, appraisal and acceptance, and research on personnel protection, etc., as well as guidance for the cockpit canopy jettisoning shock wave test experiment in this work [8]. For the impulse noise and shock wave generated when conventional weapons are launched, under the premise of considering the difference between light and heavy weapons and the number of explosions, and determined by an enormous amount of animal studies at the explosion site and army interviews, GJB2A has stipulated the method of calculating the safety limit value of the damage that may be caused to the auditory organ for the unworn protective device, and this work calculates the safety limit value of the cockpit canopy jettisoning shock wave protection according to this method, which should be the most basic protection requirement for pilots' hearing organ protection devices [9], and is fundamentally different from the protection requirement for steady state noise [10], generated by normal aircraft flight.

To study the impact of shock waves generated by aircraft cockpit emergency canopy jettisoning systems on pilot auditory organs, ground-based experiments were conducted on canopy jettisoning. Shock wave pressure data were monitored in both ears of the experimentally ejected seated dummy; combined with previous shock wave analysis methods involved in ejection rescue processes [4–7], auditory organ safety limits under canopy jettisoning shock wave conditions were calculated; the VMD-HT method was employed to perform time-frequency analysis on canopy jettisoning shock waves, aiming to design pilot auditory protection equipment effectively to improve pilot survival rates and flight safety in emergencies.

2 Canopy Jettisoning Experiment

2.1 Test Equipment

The experiment used a P3 simulation dummy as a substitute for pilots [11]. For shock wave monitoring, PCB's piezoelectric pressure sensors, model 113B27, were employed, which are IEPE type pressure sensors with a diameter of 6 mm, and the dynamic calibration of the excitation tube and the static calibration of the air pressure calibrator were completed by a third-party testing organisation prior to the test, fixed with silicone in the two ears for the simulation dummy, as shown in Fig. 1, for measuring shock wave pressures during canopy jettisoning. The data logger used was the ECHELON data logger produced by HI-TECH in the USA, supporting IEPE signal type sensors with a maximum sampling rate of 1 MHz, and sensor channel configuration, data acquisition and playback operations were completed using Aspire™ software.

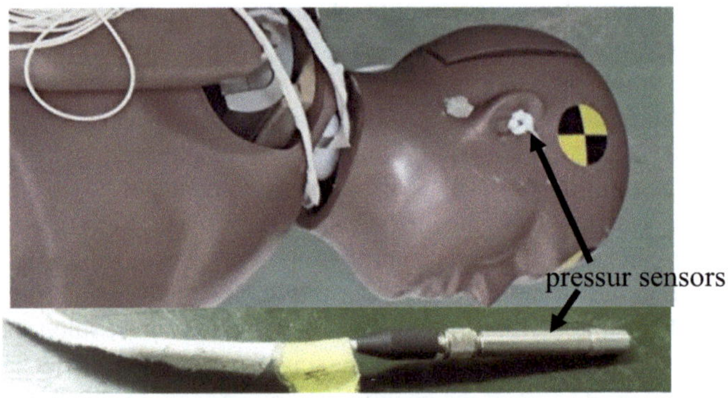

(a) Pressure sensors and the dummy

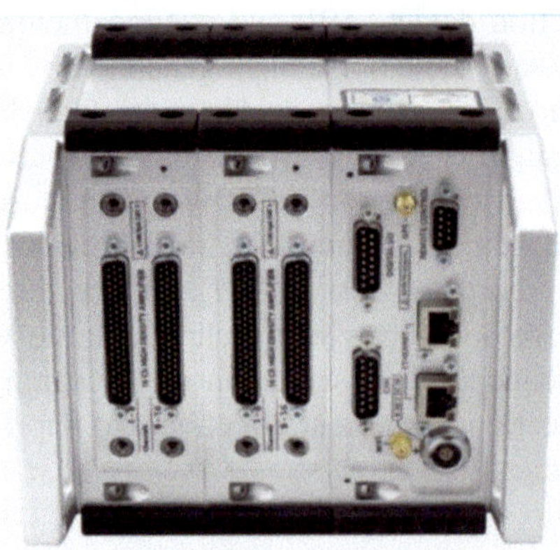

(b) Data logger (from the product manual)

Fig. 1. Test equipment

2.2 Experimental Method

To measure shock waves generated by the aircraft cockpit emergency canopy jettisoning system, the following ground test methods were established under the premise of ensuring experiment safety:

1) The test dummy wore appropriate pilot protective clothing, and the dummy was fixed on the rocket ejection seat by a restraint system, placing the dummy in an ejection state, and connecting the pressure sensor and data logger.
2) Adjust the aircraft cockpit emergency canopy jettisoning system to ensure its normal operation, then close and lock the canopy.
3) Adjust the test equipment, confirm correctness of equipment, evacuate the site, and commence the aircraft cockpit emergency canopy jettisoning test.

Fig. 2. Schematic diagram of aircraft cockpit emergency canopy jettisoning (from https://baike.baidu.com/)

3 Experimental Results

3.1 Calculation of Auditory Organ Injury Safety Limit

Through experiments, pressure-time data of shock waves of the test dummy's ears were measured. Considering the symmetry of pressure contact on both sides of the dummy's ears, analysis was performed using the example of shock waves jettisoned from the left side of the dummy, with the shock wave pressure-time curve shown in Fig. 3.

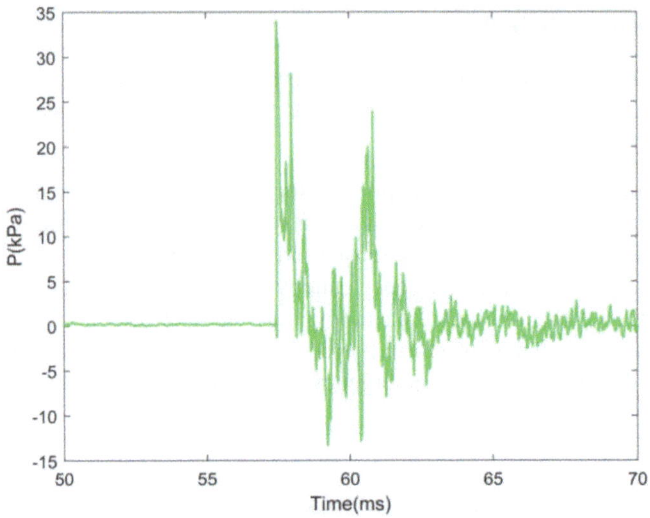

Fig. 3. Pressure-time curve of shock waves

Pressure peak: The maximum pressure peak in Fig. 2 is 34.1 kPa, converted to 184.63 dB according to (1).

$$P_{\mathrm{dB\_peak}} = 20\lg(P/P_0) \quad (1)$$

where $P_{\mathrm{dB\_peak}}$ is the corresponding decibel of positive pressure peak *P*; P_0 is the reference sound pressure level, taken as 20 μPa.

A-duration (T_{A}): The time required for pressure to increase from the surrounding atmosphere to its greatest positive pressure crest then to return to the surrounding atmosphere [9]. In this experiment, $T_{\mathrm{A}} = 0.68$ ms.

For weapons with 0.25 ms $\leq T_{\mathrm{A}} \leq$ 1.5 ms (various guns and small-caliber cannons), the safe limit for damage to personnel auditory organs due to pulse noise and shock waves is represented by (2).

$$P_{\mathrm{dB\_limit}} = 169 - 8\lg(T_A) \quad (2)$$

Substituting T_{A} into (2) yields the safety limit of 170.34 dB, hence according to [5, 10] for linear protective earmuffs, the noise reduction effect must be greater than 14.29 dB ($P_{\mathrm{dB\_peak}}$-$P_{\mathrm{dB\_limit}}$) to meet this protective requirement, the current domestic pilot helmet supporting earmuffs can meet the requirements of the protection.

3.2 Time-Frequency Characteristics Analysis of Canopy Jettisoning Shock Waves

When the VMD algorithm decomposes signals, the number of Intrinsic Mode Functions (IMF) components needs to be preset. An excessive number of decompositions may cause frequency aliasing, while too few may lose original signal characteristic information.

Therefore, this study used the central frequency similarity observation method [12] to determine that the IMF component decomposition number for canopy jettisoning shock waves was 6, with a penalty factor set at 1000 and an update rate of 0.01, which are consistent with [13].

The VMD decomposition of the canopy jettisoning shock wave pressure signal is performed. Figure 4 shows the signal that was rebuilt and the relative deviation dispersion among the rebuilt and measured signals. The figure shows that the relative deviation among the rebuilt and measured signals is less than 2×10^{-16}, indicating that the VMD method can accurately reflect the details of rapidly changing signals. This makes it suited to analyzing of shock wave signals for canopy jettisoning.

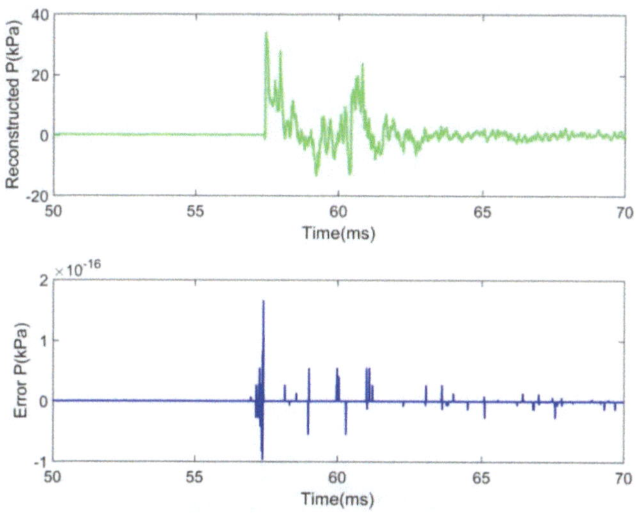

Fig. 4. VMD decomposition reconstructed signal and relative error distribution

Decomposed in order from high frequency to low frequency, six IMF components of canopy jettisoning shock wave signals obtained by VMD are shown in Fig. 5, with central frequencies [40.86, 24.75, 12.71, 5.29, 2.36, 0.41] kHz, respectively. We clearly see that IMF1 – IMF3 have the smallest amplitudes, belonging to the high-frequency component for shock wave signals, which also includes the significant quantity of white noise; the mid-frequency component is IMF4; IMF5 and IMF6 are middle and low-frequency components, with corresponding larger amplitudes, belonging to the concentrated frequency range of signal energy.

Performing Hilbert transform on each IMF component calculates the Hilbert instantaneous energy spectrum [14], shown in Fig. 6. We can find that the energy for canopy jettisoning shock waves varies with time, mainly concentrated in the interval of 57 ms to 62 ms; the predominant frequency for which is primarily located in the range of 0-5 kHz, especially with the lowest frequency energy below 1 kHz, while the energy component above 10 kHz is minimal.

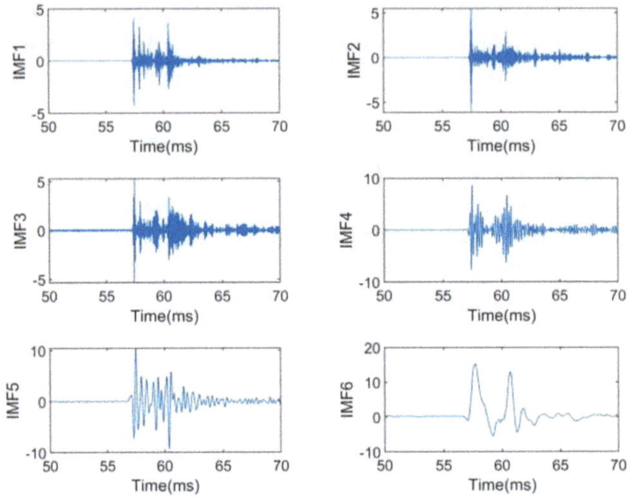

Fig. 5. VMD decomposition results

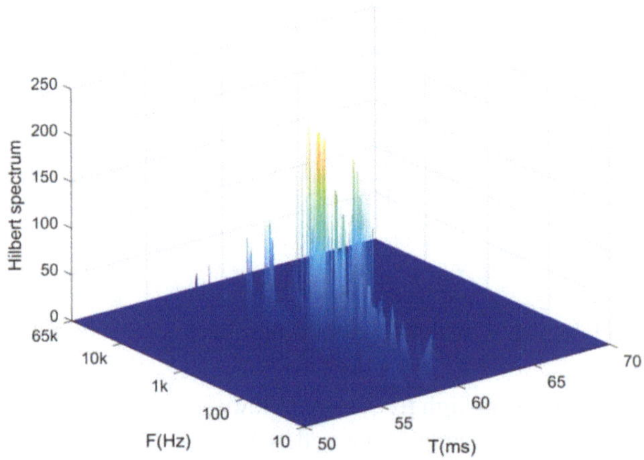

Fig. 6. Time-frequency representation of IMFs

4 Conclusion

The complex shock waves generated during aircraft cockpit canopy jettisoning pose a significant health threat to pilot auditory organs. It is essential to develop auditory protection devices adapted to the shock wave distribution characteristics. In this study, pressure sensors using piezoelectric pressure sensors in dummy ears were monitored for shock waves generated during ground-based canopy jettisoning. According to military standards, the safety limit of auditory organ damage was calculated, and preliminary protection requirements for auditory organ protection devices were proposed. The VMD algorithm and Hilbert transform were employed to analyze time-frequency characteristics of shock waves generated during canopy jettisoning, showing that shock wave energy is concentrated in low frequency region under 5 kHz, with minimal energy distribution above 10 kHz, and the duration of shock waves does not exceed 5 ms.

Currently, helmets equipped by pilots require uniform noise reduction effects within the Humans hearing frequencies range from 20 Hz to 20 kHz. However, for canopy jettisoning shock wave protection, emphasis should be placed on enhancing protection effectiveness in the low-frequency range below 5 kHz, while ensuring clear communication.

References

1. Du, J.Y., Li, J.G., Zhao, Y.N., et al.: Present situation and development trend of rocket sled ejection life - Saving test technology. Measurement & Control Technology **43**(6), 46–54 (2024)
2. Zhao, H.Q., Zhong, C.W., Han, W.C.: Numerical simulation of the influence of aircraft canopy center of gravity position on canopy jettison trajectory. Acta Aeronautica et Astronautica Sinica **29**(6), 1436–1439 (2008)
3. General Specification for Emergency Canopy Jettison System of Aircraft. HB6495 - 91
4. Guan, H., Li, G., Chen, Y., et al.: Study on the influence of explosion impact on pilots in ejection. In: Proceedings of the CSAA/IET International Conference on Aircraft Utility Systems, pp. 460–464 (2022)
5. Xie, W., Min, L., Li, J., et al.: Performance of hearing protectors in ejection seat parachute gun firing. In: Proceedings of the China Aeronautical Science and Technology Conference, Singapore: Springer Nature Singapore, pp. 49–57 (2023)
6. Li, J., Xie, W., Li, H., et al.: Time - Frequency characteristics of parachute gun shockwave based on VMD - HT. In: Proceedings of the 2023 8th International Conference on Communication, Image and Signal Processing (CCISP), IEEE, pp. 587–591 (2023)
7. Guan, H.W., Song, W.J., Zhang, X.: Research on explosion cutting impact protection technology. Aeronaut. Sci. Technol. **9**, 64–68 (2015)
8. Standard Test Method for Muzzle Blast Overpressure of Conventional Ordnance. GJB349.28 - 90, National Military Standard of the People's Republic of China (1990)
9. Safety Limit Values for Auditory Injury Caused by Pulse Noise and Blast Wave from Ordnance Firing or Explosion. GJB 2A - 96, National Defense Science and Technology Commission, Beijing (1996)
10. Kurbet, S.N., Kuppast, V.V., Chalwa, V.N.: Finite element and MBD analysis of piston to predict the engine noise. Int. J. Mech. Eng. Robot. Res. **2**(1), 183–192 (2013)
11. Jaśkiewicz, M., Frej, D., Matej, J., et al.: Analysis of the head of a simulation crash test dummy with speed motion. Energies **14**(5), 1476 (2021)
12. Liu, C., Wu, Y.: Rolling bearing fault diagnosis based on variational mode decomposition and fuzzy c means clustering. In: Proceedings of the CSEE (2015)
13. Cambow, R., Singh, M.: Comparative analysis of wear resistance for bearing coating using VMD. J. Vib. Eng. Technol. (2024)
14. Li, L., Yue, Y.: Time - frequency energy analysis of deepwater explosion shock wave signals based on HHT. In: MATEC Web of Conferences, vol. 336. EDP Sciences (2021)

Research on Statically Indeterminate Landing Gear Extension Performance

Wang Kuan(✉), Zhang Zihao, Ge Dongdong, and Xiao Peng

Shanghai Aircraft Design and Research Institute, Shanghai, China
{wangkuan,zhangzihao,gedongdong,xiaopeng}@comac.cc

Abstract. Large wide body commercial aircraft usually used the two side stays as the support and retraction & extension structure for main landing gear. The dual side stay landing gear has four attachment points with the airframe, which is a typical statically indeterminate mechanism. Compared with statically determinate structure, the dual side stay gear has to deal with the problem of extension synchronization to make a reliable extension. In this paper, taking the prototype of a dual side stay landing gear as the research object, using the rigid-flexible coupling dynamics simulation to analyse the extension and down locking performance. Focusing on the affect analysis of heading aerodynamic load and wing deformation to gear down locking performance. The lab test is also conducted to verify simulation result. The research concluded that heading aerodynamic loads can cause delays in locking the aft side stay, and with excessive heading loads even preventing the landing gear from achieving the down locking. Wing deformation can lead to the movement of the landing gear attachment points, the impact to down locking is related to the relative distance changes between after and forward side stay attachment points. The relative distance increase can help this side stay much easier to down locking.

Keywords: Statically Indeterminate Landing Gear · Dual Side Stay Landing Gear · Rigid-Flexible Dynamics Simulation

1 Introduction

Large civilian and transport aircraft experience significant loads during landing and ground taxiing due to aircraft weight. Main landing gear with two side stays is usually used by this type of aircraft, which can distribute the extra-high loads from tires to the airframe more evenly, especially for the wing made of composite material. The general geometry of the dual side stay landing gear is showed in Fig. 1.

The forward side stay and after stay are in two planes during the process of retracting and extending, which is a typical statically indeterminate mechanism. Inertia forces, self-weight, actuator forces, installation errors, aerodynamic loads, component stiffness and other factors all have a significant impact on the retracting and extending performance of landing gear [1–4]. Besides that, the effect of heading aerodynamic loads causes an overall backward deformation of the landing gear strut, resulting in tension on the forward side stay, making it easier to down locked, while the after side stay experiences

T. Zielinska (Ed.): PEAE 2025, LNNS 1679, pp. 98–106, 2026.
https://doi.org/10.1007/978-3-032-08221-3_9

compression, which is unfavourable for locking. Additionally, the wing will deformation during flight thanks to the lift force acted on the wing, this will lead to the movement of the four landing gear attachment points. This misalignment can result in a situation where one side of the landing gear lock link locks down, while the other side fails to down lock. Analyzing the impact of the aforementioned factors on the extending performance, and optimizing the corresponding structures accordingly, are the key challenges during the development of a statically indeterminate landing gear.

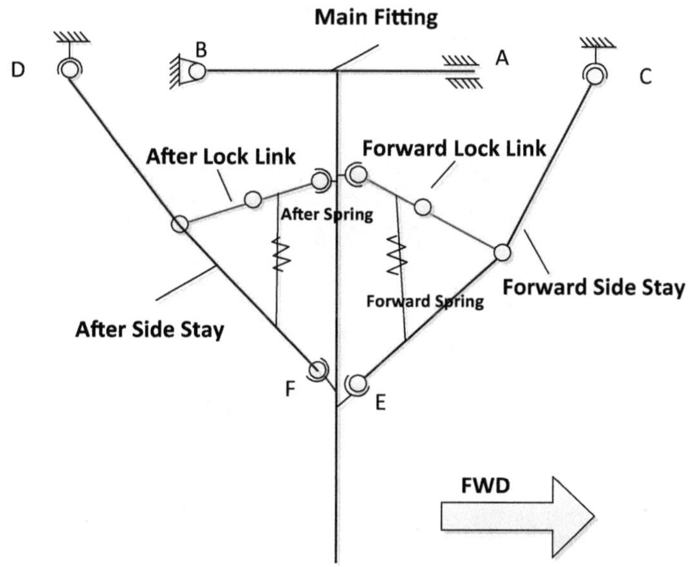

Fig. 1. Dual Side Stay Landing Gear Geometry

In this paper, by establishing a dynamics simulation model of the dual stay mechanism, the coupled analysis of mechanism dynamics, structure stiffness, heading aerodynamic loads, and attachment point movement is achieved and the general design guidelines be provided to improve the extension reliability.

2 Rigid-Flexible Coupled Dynamics Simulation

2.1 Modeling Approach and Principles

The hyperstatic characteristics of the dual side stay structure make it difficult for a purely rigid body to achieve retracting and extending analysis of the landing gear. Therefore, it is necessary to incorporate stiffness into folding components, including all the components listed in Fig. 1. In this research, finite element software is used to flexibilize those components, then import the FEM model into LMS Virtual.Lab Motion. After that, adding movable joints according to relative motion between components and setting contact commands at the lock-link locations.

Considering that folding mechanism is very complex and involves hundreds of parts, modelling it based on the actual assembly relationships would introduce numerous unnecessary constraints and makes it impossible to calculate. Additionally, it would inevitably lead to over-constraints, resulting in assembly issues with the model. Therefore, the first step is to simplify the model while ensuring the inertia and mass properties of the landing gear parts. The requirements for simplification are as follows: maintaining consistent relative motion relationships with actual motion conditions and aligning mass characteristics with physical samples. Based on the requirements for model simplification, analyse the mutual motion relationships between each component and simplify the landing gear model accordingly. The main landing gear rigid-flexible model is shown in Fig. 2.

Fig. 2. Landing Gear Extension Rigid-Flexible Model

To achieve the attachment points movement, the four attachment pins or cardans are connected to four virtual attachment structure by "Cylinder Joint", and then the four virtual attachment structure are connected with the fixed ground by "Distance Constraint". By this way, the attachment position can be changed by adjusting the "Distance Constraint" value.

2.2 Model Properties

The main properties of the dynamic model including the weight, inertia, lock spring stiffness and component stiffness. The main properties are shown in Table 1.

Table 1. Main Components Properties.

Property	Value
Main fitting weight (kg)	1966.07
AFT side stay weight (kg)	137.94
Lock link weight (kg)	18.33
FWD side stay weight (kg)	164.3
Bogie beam assembly weight (kg)	677.1
Axle assembly weight (kg)	210.16
Bogie pitch trimmer weight (kg)	18.56
Wheel &Tire weight (kg)	349.6
Lock Spring Stiffness (N/mm)	6

2.3 Simulation Analysis

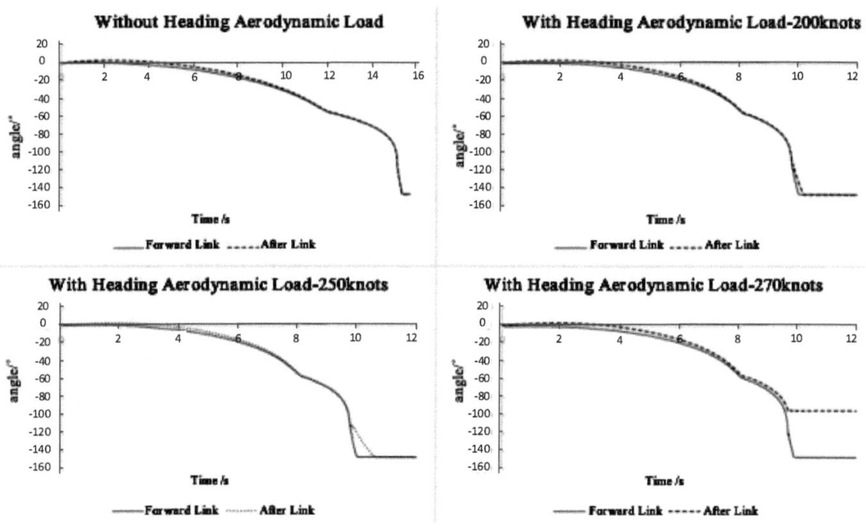

Fig. 3. Unfolding Angle of Lock Link

Simulation focused on the deflections due to movement of the attachment points (thanks to wing twist and bend) and aerodynamic heading loads applied to the landing gear.

Heading aerodynamic loads can cause the main fitting of the landing gear to bend backward, resulting in compression of the after side stay and shortening the distance between points D and F (showed in Fig. 1). On the other side, the forward side brace is stretched, extending the distance between points C and E. Consequently, this scenario prioritizes the locking of the forward lock link over the after lock link. When the heading aerodynamic loads are excessive to about 100 KN, it may prevent the after lock link from reaching the locked state, as shown in Fig. 3.

For the wing deformation impact. If there is a significant deviation from the attachment points, it will lead to the landing gear can't fully down locked. The simulation analysis results of attachment points deviating from its original position are shown in the following Table 2.

After analyzing the simulation results, the deformation of the wing leading to the inability of locking complies the following principles:

- Taking the four attachment position of the landing gear installation on the aircraft assembly line as the theoretical position. Define the distance between the upper and lower attachment of the forward side stay (distance between C and E) is l_1, and the distance between the upper and lower attachment of the after side stay (distance between D and F) is l_2. The lengths then deformed during in flight to l_1' and l_2';
- Define $\Delta l_1 = l_1 - l_1'$, $\Delta l_2 = l_2 - l_2'$ and $\Delta l = \Delta l_2 - \Delta l_1$. On the assembly line, both Δl_1, Δl_2 and Δl are zero;
- When $\Delta l < a$ ($a < 0$, a is the lower limit, which is related to the structural stiffness), the relative position of the after side stay is compressed, and the after side stay tend to unable to down locked. If $\Delta l > b$ ($b > 0$, b is the upper limit value, which is related to the structural stiffness), then the relative position of the forward side stay is compressed, and the forward side strut tend to unable to down locked.

Table 2. Wing Deformation Simulation Results

Case	Offset of Point C/mm			Offset of Point D /mm			Down-lock Result
	H	V	L	H	V	L	
C1	+15	0	0	0	0	0	AFT lock link can't down locked
C2	−15	0	0	0	0	0	FWD lock link can't down locked
C3	0	+15	0	0	0	0	AFT lock link can't down locked
C4	0	−15	0	0	0	0	FWD lock link can't down locked
C5	0	0	+15	0	0	0	AFT lock link can't down locked
C6	0	0	−15	0	0	0	FWD lock link can't down locked
C7	0	0	0	+15	0	0	FWD lock link can't down locked
C8	0	0	0	−15	0	0	AFT lock link can't down locked
C9	0	0	0	0	+15	0	FWD lock link can't down locked
C10	0	0	0	0	−15	0	AFT lock link can't down locked
C11	0	0	0	0	0	+15	FWD lock link can't down locked
C12	0	0	0	0	0	−15	AFT lock link can't down locked

The above mentioned are two outer factor that will affect the down locking reliability, there are two inner factor will also help improving the down locking performance.

The first one is down locking spring. The spring force can help the lock link to pass the unlocked position. In the simulation, take C12 in Table 2 as the an example, it can't achieve the after lock link locking under spring stiffness of 6 N/mm. The stiffness of the spring is then tuned to 8 N/mm, 10 N/mm and 12 N/mm, and the locking synchronization of both sides are showed in Table 3 which is the key symbol of locking performance. The result showed that higher spring stiffness can help to increase the downlocking performance, but the effect will be diminished after higher than certain value and the weight will also increase with the stiffness increased.

Table 3. Down Locking Performance With Different Spring Stiffness

Spring Stiffness/ N/mm	Time Delta/s
6	∞
8	0.18
10	0.06
12	0.05

For the components' stiffness impact, three different side stay materials including aluminum, titanium and steel are simulated under certain heading aerodynamic load, as is shown in Table 4. The result shows that different stiffness have limited affect for down locking performance. The reason is softer component will be easier to pass the locking position under heading aerodynamic load, but harder component will help to minimize the structure deformation, which will help to down locked.

Table 4. Down Locking Performance With Different Component Stiffness.

Material	Time Delta/s
aluminum	0.05
titanium	−0.05
steel	−0.08

3 Test Verification

3.1 Test Preparation

The lab test with a full scale specimen is conducted to verify the result of the dynamic simulation, especially for the impact of heading aerodynamic load and wing deformation. The specimen is installed with the rigid test rig as shown in Fig. 4.

Fig. 4. Test Rig and Installation

3.2 Heading Aerodynamic Load Impact Test

The heading load is loaded by a steel wire, which one end is connected with specimen's main fitting component and the other end is connected with the loading system, as is shown in Fig. 5.

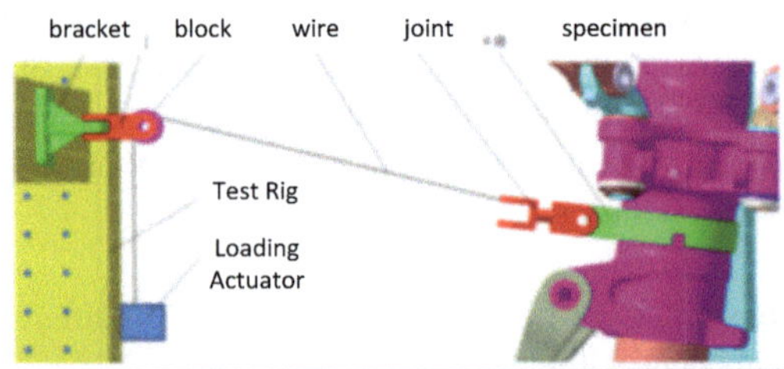

Fig. 5. Heading Aerodynamic Loading System

To validate the heading aerodynamic load impact, six sets of tests were conducted with loading forces ranging from 10 KN to 105 KN to simulate the loads generated under different aircraft airspeed. The test results are as follows:

- Under low load conditions (10 KN to 30 KN), the landing gear can be extended and down locked both under alternate extension (which only by structure' own weight) and hydraulic pressured extension;
- Under moderate heading load conditions, the after strut lock link cannot be locked under gravity conditions, but can be locked with the help of hydraulic pressure. And as the heading load increases, the hydraulic pressure required for locking also increases;
- Under the simulated airspeed condition of 250 km/h with a 105 KN load, the after side lock link can't be locked even under hydraulic pressured.

Through the tests, it can be determined that after the implementation of heading aerodynamic loads, the landing gear shock absorber strut as a whole undergoes a slight backward deformation and tilt, and the distance between the upper and lower attachment points of the after side stay decreases, leading to difficulties in locking. The test results validate the effectiveness of simulation result.

3.3 Wing Deformation Impact Test

To validate the wing deformation impact, the forward side stays and after side stay upper attachment joints are moveable, as showed in Fig. 6. Constrained by the difficulty of technical implementation, the move can only along upper-down and spanwise directions.

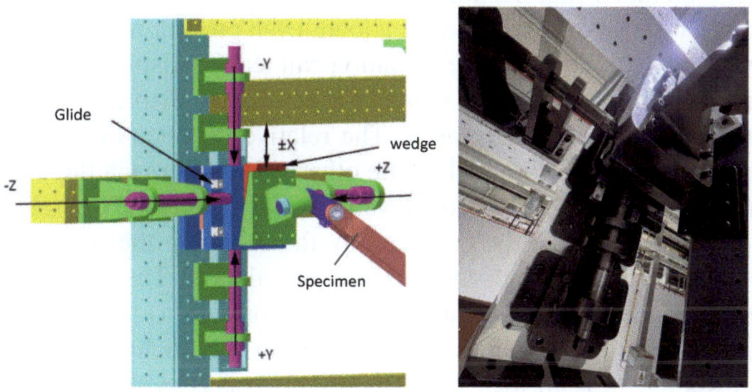

Fig. 6. Wing Deformation System

Although only two of the four attachment points are moveable, it is enough to simulate the relative distance of the two side stays. During the test, different offset was conducted from small to large, and the milestone was listed in Table 5.

Table 5. Wing Deformation Test Results

Case	Offset of Point C/mm			Offset of Point D /mm			Down-lock Result
	H	V	L	H	V	L	
S4	0	−25.0	0	0	0	0	FWD lock link can't down locked
S9	0	0	0	0	35.2	0	AFT lock link can't down locked
S12	0	0	0	0	0	14.6	AFT lock link can't down locked

The test results are more optimistic with larger offset millstone compared with the simulation results. The reason is the joints gap, spherical bearings and test rig stiffness in the lab test will counteract the impact of the attachment move. However, the principle that relative distance increase can help this side stay much easier to down locking can still be validated through the test.

4 Conclusion

In this paper, extension performance simulation, design and lab test method of the statically indeterminate landing gear with dual side stays are discussed. The simulation model using rigid-flexible coupling dynamics simulation to analyse the outer factors including heading aerodynamic load and wing deformation and landing gear structure factors including down lock spring stiffness and components' stiffness. The lab tests provide an implementable way to test full scale landing gear structure under heading aerodynamic load and wing deformation and to validate the simulation results.

The paper concluded that heading aerodynamic loads can cause delays in locking the aft side stay, and with excessive heading loads even preventing the landing gear from achieving the down locking. Wing deformation can lead to displacement of the landing gear attachment points, the effect is related to the relative displacement between after and forward side stay attachment points. The relative distance increase can help side stay much easier to down locking, while the other side can't reach the locked position.

Based on the simulation and test conclusion, the design method of a reliable extension dual side stay mechanism can be achieved by firstly sizing the mechanism and then validated the key factors through the dynamic simulation. The sizing process has to consider the heading aerodynamic load and wing deformation effect separately and also their combined effect. The possible ways to achieve this include stronger down lock springs and lower stiffness aft lock stay. The actual properties should be determined by dynamics simulation.

References

1. Rui, F.: Analysis and test for retraction-extension dynamic performance of a civil aircraft main landing gear. Nanjing University of Aeronautics and Astronautics, Nanjing (2012)
2. ZhiJuan, S., JingTao, D.: Research on kinematic and dynamic simulation of landing gear retraction mechanism in aircraft. Manuf. Autom. **13**, 97–100 (2014)
3. Yin, Y.: Dynamics and reliability analysis of retraction of landing gear. Nanjing University of Aeronautics and Astronautics, Nanjing (2012)
4. Knowles, J.A.C., Lowenberg, M.H., Neild, S.A., et al.: A bifurcation study to guide the design of a landing gear with a combined uplock/downlock mechanism. Proc. R. Soc. A Math. Phys. Eng. Sci. **470**(2172), 1–22 (2014)
5. Knowles, J.A.C.: Bifurcation study of a dynamic model of a landing-gear mechanism. J. Aircr. **53**(5), 1468–1477 (2016)
6. Yin, Y., Hong, N., Xiaohui, W.: Retraction system performance analysis of landing gear with the influence of multiple factors. J. Beijing Univ. Aeronaut. Astronaut. **41**(5), 953–960 (2015)
7. Yi, T., Xiang, X., Nan, L.: Computer analysis of large scale aircraft landing gear retraction and extension control system. J. Beijing Univ. Aeronaut. Astronaut. **39**(5), 596–599 (2013)

Study on the Vibration of the Auxiliary Power Unit and Cabin Noise in Civil Aircraft

Han Wang(✉), Jingtian Sha, Jinhua Lyu, Xiaolong Li, and Jiafeng Tong

Shanghai Aircraft Design and Research Institute, Shanghai, China
wanghan1@comac.cc

Abstract. The Auxiliary Power Unit (APU) is a widely used onboard device in civil aircraft and is considered one of the sources of cabin noise. Understanding the relationship between APU vibrations and the cabin noise helps in designing better vibration isolation systems. This paper presents tests conducted on the APU vibration and cabin noise under different operating conditions for a tail-mounted engine civil aircraft. The results indicate that when the APU generator is operating, there is a significant increase in vibration energy at the generator rotor frequency. Additionally, when the APU provides bleed air, the vibration energy at the APU rotor frequency also increases. The highest average sound pressure level in the rear cabin was observed when the APU was simultaneously generating power and providing bleed air. There is a strong correlation between the cabin noise sound pressure levels at the APU generator frequency and rotor frequency and the vibration energy of the APU itself. Specifically, during APU electrical power generation, the vibration energy at the generator rotor frequency significantly increases, leading to a marked increase in the cabin sound pressure level at that frequency. Therefore, when designing the APU vibration isolation system, it is important to focus on effectively reducing the transmission of vibrations and noise from the APU generator and APU main shaft rotor frequencies.

Keywords: Civil Aircraft · Auxiliary Power Unit · Vibration · Cabin Noise · Ground Test

1 Introduction

The Auxiliary Power Unit (APU) is a widely used airborne equipment for civil aircraft. The primary functions of the APU in civil aircraft are to provide bleed air for engine start and environmental control system, as well as to generate electrical power by driving a generator [1, 2]. On the ground, before the main engines are started, the APU is usually responsible for supplying the aircraft with electrical power and air. Generally, the APU is a small gas turbine engine, and the rotation of its components such as the compressor, turbine, and gearbox generator generates vibrations. These vibrations inevitably produce noise [3], thus making the APU one of the sources of aircraft cabin noise [4].

There are many studies on using APU noise for fault diagnosis and health monitoring [5–7], However, there are few reports on the impact of APU vibration noise on the cabin. Many factors could influence cabin noise. Typically, when the aircraft is on the ground,

T. Zielinska (Ed.): PEAE 2025, LNNS 1679, pp. 107–115, 2026.
https://doi.org/10.1007/978-3-032-08221-3_10

the operation of the APU, environmental control system, hydraulic systems, and various electrical devices all contribute to cabin noise [8]. Many civil aircraft install vibration isolators or vibration absorbers at the APU mounting points to reduce the transmission of vibrations and noise between the APU and the aircraft structure [9, 10]. This enhances the comfort and safety of the aircraft, thereby improving its competitiveness. Therefore, studying the relationship between APU vibrations and cabin noise helps in designing better APU vibration isolation systems.

This paper focuses on the different ground usage scenarios of the APU and conducts tests on APU vibration and cabin noise for a tail-mounted engine civil aircraft. The influence of different APU conditions on cabin noise is analyzed, and the correlation between APU vibration and cabin noise is studied. This paper providing references for APU vibration reduction and cabin noise reduction design of civil aircraft.

2 APU Vibration and Cabin Noise Testing

2.1 Testing Equipment and Measurement Points

The tests were carried out on a civil aircraft with a tail-mounted main engine. The APU is installed in the APU compartment at the rear of the aircraft using seven tie rods and three mounting points, with a vibration isolator installed at each mounting point. Figure 1 shows the position of the APU on the aircraft and provides a schematic of the aircraft's layout.

For cabin noise measurement, Fifteen EM46AE fixed noise sensors (microphone) with a sampling frequency of 65536 Hz were placed at different locations in the rear of the cabin. The noise sensors were positioned at seat height to simulate the area around human ears, as shown in Fig. 1. One triaxial 7253DM3–10 vibration sensor with a sampling frequency of 8000 Hz was placed on the APU gearbox to measure the vibration of APU, also as shown in Fig. 1.

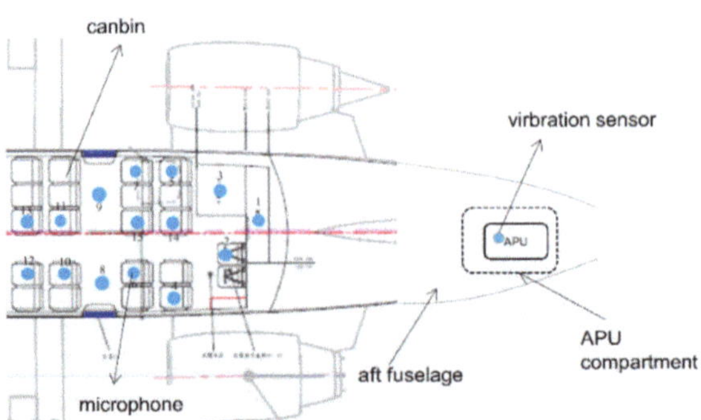

Fig. 1. Sample Schematic diagram of the noise sensors in the rear cabin and the APU vibration sensor measurement points

2.2 Testing Conditions

Typical ground usage scenarios and conditions for the APU include the following: APU running at no load after startup (APU NL), APU providing bleed air for the environmental control system (APU BL), APU generating electrical power for the aircraft (APU GEN), and APU simultaneously providing electrical power and bleed air (APU GEN&BL). Vibration data and noise data was collected under above APU operating conditions. Additionally, considering the aircraft's main engines are tail-mounted, cabin noise data was also collected with both engines idling and the APU shut down for comparison purposes.

2.3 Data Processing

The vibration acceleration data was processed to obtain the power spectral density (PSD) in the frequency range of 20 Hz to 3.2 kHz by software MATLAB. A Hanning window was applied, and the results were displayed in decibels (dB). One-third octave band analysis was performed on the noise data of different measurement points using B&K Reflex software. The A-weighted sound pressure levels were calculated, and the average and maximum values were analyzed. Additionally, for the measurement point with the highest noise level, the PSD was obtained through software data processing in the frequency range of 20 Hz to 3.2 kHz, using linear averaging and a Hanning window.

3 Results and Discussion

3.1 APU Vibration Under Different Operating Conditions

A triaxial accelerometer was installed on the APU gearbox, and the results for the three directions under different conditions showed similar patterns. One direction was selected as a representative for analysis, as shown in Fig. 2. It can be observed that the APU vibration power spectrum is closely related to the rotor frequencies, where fg represents the rotor frequency of the generator mounted on the APU, and fc represents the frequency of the APU main shaft rotor. When the APU is in no-load and bleed air modes, i.e., APU NL and APU BL, the harmonic components of the APU generator rotor and APU main shaft rotor are not prominent, indicated by fewer peak values in the spectrum. However, when the generator is supplying electrical power, i.e., in APU GEN and APU GEN&BL modes, the number of harmonic components significantly increases, resulting in more peak values in the spectrum.

Figure 3 is a partial enlargement of the frequencies f_g and f_c. It shows that when the APU generator is supplying electrical power, the vibration energy at the generator rotor frequency f_g significantly increases. The values for APU GEN and APU GEN&BL are approximately 17 dB higher than those for APU NL and APU BL at f_g frequency, as shown in Fig. 3a. The increase in the number of peaks in the vibration power spectrum and the increase in vibration energy at specific frequencies indicate a significant overall increase in APU vibration energy when the generator is supplying electrical power. The APU generator is typically mounted on the gearbox, which contains many gears and other mechanical components. During operation, gear meshing, clearances, and friction can lead to vibrations, which are exacerbated under load, potentially increasing the overall vibration energy [11]. As a result, in the design of the APU vibration reduction system, reducing the transmission of vibrations from the APU generator to the aircraft is a critical focus.

Additionally, Fig. 3b shows that at the APU main shaft rotor frequency fc, the vibration energy is higher in APU BL and APU GEN&BL modes compared to APU NL and APU GEN modes. When the APU is in bleed air mode, the compressor load increases, resulting in higher vibration energy at the APU main shaft rotor frequency. Typically, APU rotor impellers are finely balanced due to their high rotational speeds, resulting in minimal imbalance. However, as the air load increases, the rotor frequency vibration energy increases, although not as significantly as the generator rotor. Hence, vibration attenuation at the APU main shaft frequency should be included in the design considerations for the APU vibration reduction system.

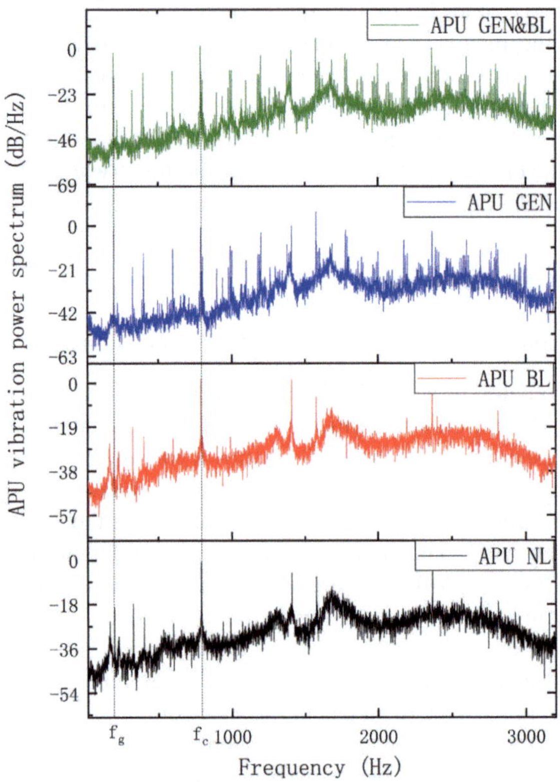

Fig. 2. Vibration Power Spectrum of APU Gearbox

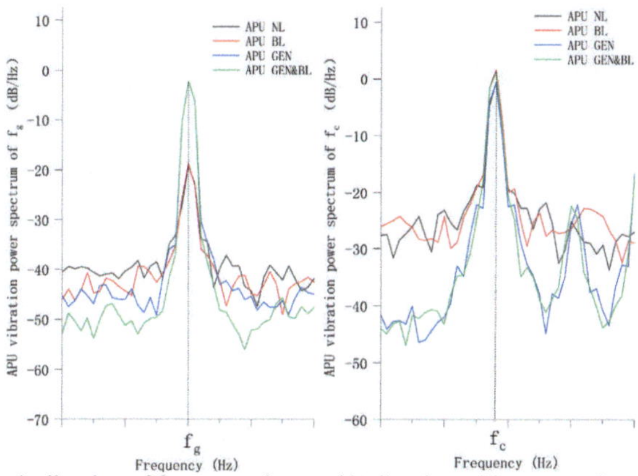

a) vibration of frequency fg b) vibration of frequency fc

Fig. 3. Enlarged Vibration Power Spectrum at APU Generator Rotor Frequency f_g and APU Main Shaft Rotor Frequency f_c

3.2 Cabin Noise Data Analysis

The average and maximum noise levels at the rear of the cabin under different APU operating conditions are shown in Fig. 4. It can be seen that for the four typical APU operating conditions, the average noise levels in the rear cabin are ranked as follows: APU NL < APU BL < APU GEN < APU GEN&BL. Compared to APU NL, in APU BL mode, besides the APU providing bleed air, the aircraft's environmental control system equipment also starts working, which contributes to the overall noise increase. Despite this, APU BL and APU NL have similar vibration energy levels, leading to comparable maximum noise levels.

In APU GEN mode, the noise level is higher than in APU BL mode. This is partly due to the significant increase in APU vibration energy mentioned above. Additionally, electrical systems such as the hydraulic system also start working in APU GEN mode, contributing to the cabin noise. The average noise level in APU GEN&BL mode is the highest among the four conditions, due to both increased APU vibration and the operation of environmental control system and hydraulic systems. However, the maximum noise levels in APU GEN and APU GEN&BL modes are comparable, suggesting that the generator's contribution to noise is more significant in APU GEN&BL mode.

For comparison, cabin noise was also tested with both main engines at idle and the APU shut down. The rear cabin noise level in this scenario is significantly higher than in any APU operating condition. The average noise level in APU NL mode is nearly 14 dBA lower than with both engines idling. The aircraft used in this test has rear-mounted engines rigidly connected to the fuselage, allowing engine noise to be directly transmitted to the cabin, resulting in higher noise levels. In contrast, the APU is located in the tail cone and connected to the fuselage through vibration isolators, resulting in significantly lower noise levels in APU NL mode compared to both engines idling.

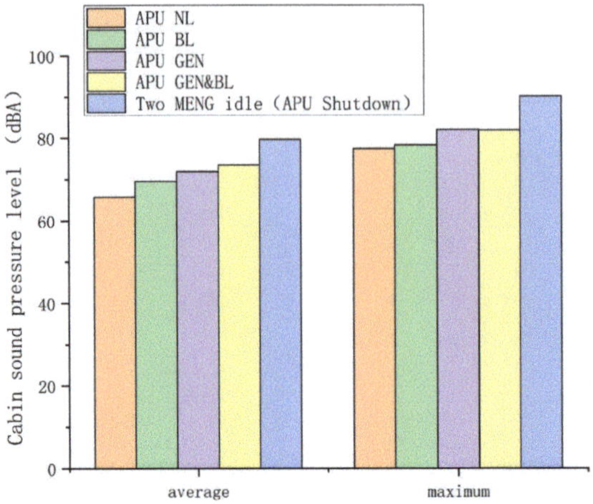

Fig. 4. Sound Pressure Levels in the Rear Cabin under Different Conditions

The maximum sound pressure levels under different APU operating conditions were all measured at the same point (above the fourth row of seats at the rear). The sound power spectrum at this point is shown in Fig. 5. The APU rotor frequency fc is clearly visible in all conditions. However, the generator rotor frequency fg is not prominent in APU NL and APU BL modes, but shows a clear peak in APU GEN and APU GEN&BL modes. This peak is the highest across the frequency range, with additional peaks appearing at 2*fg, 3*fg, etc. (harmonic frequencies of the generator rotor), consistent with the vibration test results. Peaks at ~1200 Hz in APU GEN and APU GEN&BL modes correspond to the working frequency of the hydraulic system, while peaks at ~1900 Hz and ~2100 Hz in APU BL and APU GEN&BL modes are related to the environmental control system.

The increased vibration energy at the generator rotor frequency fg when the APU generator is loaded results in a corresponding increase in noise at this frequency. Figure 6a shows an enlarged view at fg, revealing a strong correlation between sound pressure levels and vibration energy at this frequency, as discussed before. Figure 6b shows an enlarged view at fc, at the APU main shaft rotor frequency fc, sound pressure levels in APU BL and APU GEN&BL modes are higher than in APU GEN and APU NL modes, consistent with the vibration test results. However, despite similar vibration energy levels, the sound pressure level at fc in APU BL mode is higher than in APU GEN&BL mode. This may be due to attenuation of the APU main shaft rotor vibration as it transmits to the gearbox, resulting in less pronounced increase in vibration energy at the fc frequency for the vibration sensors on the gearbox.

Overall, there is a strong correlation between vibration and noise data at the rotor frequencies of the APU generator and APU main shaft. Therefore, when designing the APU vibration isolation system, it is necessary to comprehensively consider both vibration and noise reduction aspects when setting performance requirements such as transmissibility of vibration isolation system. Additionally, the focus should be on effectively reducing the transmission of vibrations and noise from the APU generator and APU main shaft rotor frequencies.

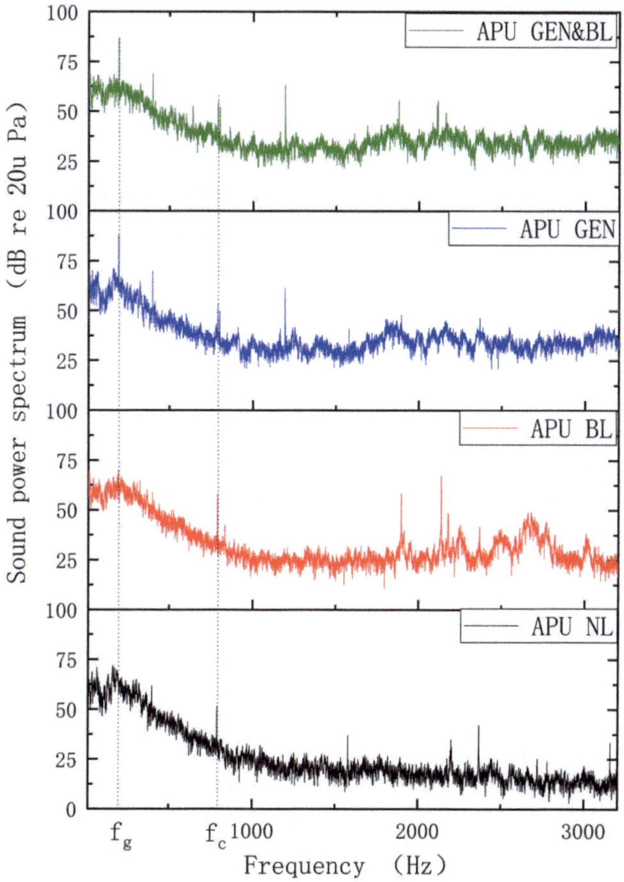

Fig. 5. Sound Power Spectrum in the Cabin under Different APU Operating Conditions

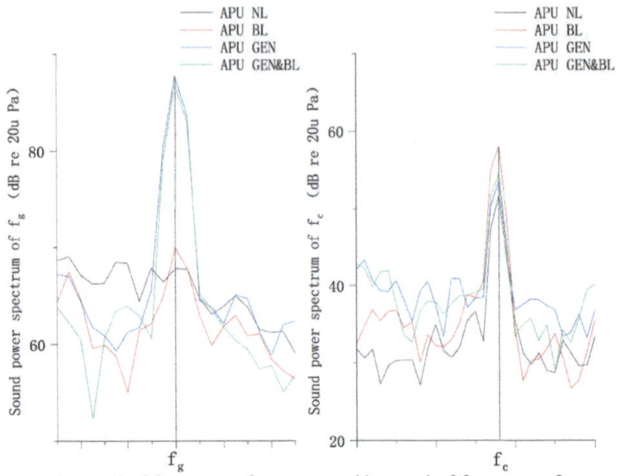

a) sound of frequency fg b) sound of frequency fc

Fig. 6. Enlarged Sound Power Spectrum at APU Generator Rotor Frequency f_g and APU Main Shaft Rotor Frequency f_c

4 Conclusion

The Auxiliary Power Unit (APU) is a widely used onboard device in civilian aircraft. While the APU is in operation, vibrations from components such as rotating parts produce noise that can be transmitted to the cabin. When the APU generator is operating, the load on the motor exacerbates the vibration of the various rotors in the gearbox, significantly increasing the vibration energy at the generator rotor frequency. Additionally, harmonic frequencies across the entire frequency band become more pronounced. When the APU is providing bleed air, the increased load on the compressor rotors results in higher vibration energy at the APU rotor frequency.

Under different typical APU operating conditions, A-weighted sound pressure levels indicate that the highest levels occur when the APU simultaneously generates electrical power and provides bleed air, followed by scenarios where the APU generates electrical power, provides bleed air, and operates in no-load mode. The cabin noise levels at the APU generator and rotor frequencies show a strong correlation with the APU's vibration energy. When the APU is generating electrical power, the increased vibration energy at the generator rotor frequency results in significantly higher sound pressure levels at this frequency. Similarly, increased vibration energy at the APU rotor frequency also leads to higher sound pressure levels in the cabin at this frequency. Therefore, when designing the APU vibration isolation system, it is important to focus on effectively reducing the transmission of vibrations and noise from the APU generator and APU main shaft rotor frequencies.

References

1. Li, D.: Application status and development trend of auxiliary power unit. Aeronaut. Sci. Technol. **000**(006), 7–10 (2012)

2. Li, B., Tang, H., Yin, H., et al.: APU mount system design and analysis. Civil Aircraft Des. Res. **03**, 42–46 (2010)
3. Ahmed, U., Ali, F., Jennions, I.K.: Evaluation of aircraft auxiliary power unit near-field acoustics for condition monitoring. IEEE Access **10**, 108564–108582 (2022)
4. Sun, X.: The study of commercial aircraft cabin noise resources test. Sci. Technol. Vis. **14**, 88–89 (2014)
5. Ahmed, U., Ali, F., Jennions, I.: Acoustic monitoring of an aircraft auxiliary power unit. ISA Trans. **137**, 670–691 (2023)
6. Ahmed, U., Ali, F., Jennions, I.: A review of aircraft auxiliary power unit faults, diagnostics and acoustic measurements. Prog. Aerosp. Sci. **124**, 100721 (2021)
7. Skliros, C., Ali, F., Jennions, I.: Fault simulations and diagnostics for a Boeing 747 auxiliary power unit. Expert Syst. Appl. **184**, 115504 (2021)
8. Hu, X., Han, F., He, L., et al.: Overview of the cabin acoustic design for the large commercial aircraft (I). Civil Aircraft Des. Res. **02**, 1–3 (2011)
9. Huang, F., Yao, P., Miu, Z.: Design of strut suspended APU mounting system. Aeronaut. Sci. Technol. **28**(06), 26–29 (2017)
10. Tang, Z., Luo, G., Chen, W., et al.: Dynamic characteristics of vibration system including rubber isolator. J. Nanjing Univ. Aeronaut. Astronaut. **46**(02), 285–291 (2014)
11. Kamde, L.G., Patil, S.A., Gujar, U.N.: Identification of acoustic emission source and comparative study on the diagnosis of a spur gearbox. Int. J. Adv. Res. Sci. Eng. Technol. **01**(04), 01–07 (2012)

An Indoor Cargo Deliver System Using Drone and Magnetic Catcher

Mingyang Liang[1], Qihang Zhou[2], Yafeng Xue[3], Yuan Zhao[4], and Long Chen[5(✉)]

[1] Shenzhen Institute of Artificial Intelligence and Robotics for Society, Shenzhen 518000, China

[2] The School of Mechanical Engineering, Northwestern Polytechnical University, Xi'an 710000, China

[3] The School of Cybersecurity, Northwestern Polytechnical University, Xi'an 710000, China
xueyafeng@mail.nwpu.edu.cn

[4] Changan Area Distribution Network Center, Stage Grid Xian Electric Power Supply, Xi'an 710000, China

[5] The School of Mechanical Engineering, Northwestern Polytechnical University, Xi'an 710000, China
chenmf767@foxmail.com

Abstract. This paper primarily developed and tested a drone cargo delivering system in an indoor environment. We use motion capture to address the localization issue. Through a sophistic design on flight routine, the deliver drone was able to autonomously complete high-precision cargo transportation and round-trip takeoff and landing tasks without any manual intervention, with minimal error margins. This work provides a valuable forward-looking study for the commercial application of drone application in indoor settings. And we will further extend this result to the city environment.

Keywords: Robotics · drone · motion capture · cargo delivery

1 Introduction

For the development of the drone industry, cargo delivering systems for drones have long been a challenging issue [1]. Moreover, the limited indoor environment, changing lighting conditions, and the presence of both dynamic and static obstacles pose significant challenges to the positioning and navigation systems of drones. Compared to outdoor navigation tasks, indoor missions require drones to have higher positioning accuracy, more reliable mapping capabilities, and greater autonomy [2]. Interestingly, drones can effectively utilize aerial space indoors, allowing for flexible and direct cargo transportation in confined areas, serving as a strong complement to ground-based and manual handling [3]. Indoor drone delivering typically employs either suspended cargo or direct attachment methods for transport. Different cargo attachment methods also entail different cargo pickup techniques. Additionally, the use of cable suspension is highly influenced by the environment; the cable length must not be too long, and the weight of the cargo significantly impacts the drone's flight stability [4, 5].

T. Zielinska (Ed.): PEAE 2025, LNNS 1679, pp. 116–126, 2026.
https://doi.org/10.1007/978-3-032-08221-3_11

Due to the inherently small size of drones, their load-carrying and cargo grasping capabilities are quite limited. Most researches have focused on precise navigation and flexible adaptive mechanisms. However, there are also studies that have significantly enhanced the load-bearing capabilities of drones using new bio-inspired technologies [6, 7].

To enable the drone to autonomously complete the entire delivering process and investigate the impact of environmental factors, this paper primarily introduces a drone cargo delivering system under ROS [8] environment, which has been tested in multiple indoor scenarios. The system operates without any manual intervention and meets the precision and reliability requirements for cargo delivering tasks. Sect. 2 focuses on the components of the drone cargo delivering system, while Sect. 3 provides a detailed description of its workflow and algorithms. Sect. 4 presents the real-world delivering tests conducted in indoor scenarios, and the paper concludes with the final findings and implications.

2 Components of the Drone System

2.1 System Architecture

As is shown in Fig. 1, the motion capture identifies the reflective markers attached to the drone, obtaining precise positional information for each Markers. This data is processed by the ground-based positioning computer (GPC) to calculate the drone's real-time position. The drone's onboard computer subscribes to VRPN (Virtual-Reality Peripheral Network) ROS topic messages to receive this positioning data. Meanwhile, the GPC uses the ROS framework to communicate with the drone's onboard computer (DOC) via ROS. Within the coordinate system of the motion capture, the GPC publishes the coordinates of the target anchor points and issues flight commands to achieve precise drone navigation. The DOC processes the received target coordinates and flight commands, converting ROS messages into Mavlink signals via the Mavros middleware, and transmits them to the flight controller PX4 for execution.

Additionally, the drone is equipped with a pod. Upon reaching the target anchor point, the GPC uses the ROS Toolbox to send a cargo release ROS topic message to the ESP32 chip on the pod, while also sending a waiting-to-receive ROS topic message to the ESP32 chip on the ground winch designed to catch the cargo. The ESP32 chips process these ROS topics and use the ros-serial library for serial communication to control the pod release and winch motor operation. When the winch detects the cargo drop, it automatically secures the cargo and sends a cargo-received ROS topic message back to the GPC. At this point, the GPC sends commands to the DOC and flight controller to instruct the drone to return and land.

Throughout the entire task process, all steps are executed automatically by the program, with the remote control only used to ensure safety in case of unexpected situations. The detailed components of the system are shown in the Table 1.

Table 1. The detailed components of the system

Drone	PX4 Flight Controller	Manages the drone's flight dynamics and stability by executing the Mavlink commands received through the Mavros middleware.
	Onboard Computer (DOC)	Acts as the central processing unit for the drone, handling ROS (Robot Operating System) communication. It subscribes to VRPN topics to acquire precise positional data and processes commands from the control computer.
	Nacelle	Contains an ESP32 chip to manage the cargo release mechanism, receiving commands via ROS and converting them into serial signals for execution.
Motion Capture	Infrared Cameras	Identify reflective markers attached to the drone, providing precise positional data for each marker.
	Switch and Router	Facilitate data transmission from the motion capture cameras to the ground-based positioning and control computer.
Ground-Based Systems	Positioning and Control Computer (GPC):	Utilizes ROS framework to communicate with the DOC. It publishes target anchor coordinates and flight commands within the motion capture's coordinate frame. Additionally, it controls the cargo release and receiving mechanisms by sending ROS topic messages to the respective ESP32 chips.
	Catcher (Capstan) [9]	Equipped with an ESP32 chip to manage the winch mechanism for receiving cargo. It processes ROS topics via the ros-serial library and controls the winch motor to secure the dropped cargo.
Communication and Control Flow	VRPN (Virtual-Reality Peripheral Network)	Used for transmitting precise positional data from the motion capture to the DOC.
	ROS	Facilitates communication between the GPC and the DOC, as well as between the GPC and the ESP32 chips on the nacelle and capstan.
	Mavros Middleware	Converts ROS messages into Mavlink signals, which are then executed by the PX4 flight controller.
	Serial Communication	Used by the ESP32 chips to control the nacelle and capstan mechanisms based on the received ROS topic messages.
Additional Components	Remote Control	Primarily serves as a backup to ensure safety during unexpected situations, allowing manual intervention if necessary.
Data and Signal Flow	Motion Capture Data	Transmitted from the cameras to the GPC.
	Control Signals	Sent from the GPC to the drone and cargo handling mechanisms.

(*continued*)

Table 1. *(continued)*

Drone	PX4 Flight Controller	Manages the drone's flight dynamics and stability by executing the Mavlink commands received through the Mavros middleware.
	Onboard Computer (DOC)	Acts as the central processing unit for the drone, handling ROS (Robot Operating System) communication. It subscribes to VRPN topics to acquire precise positional data and processes commands from the control computer.
	Nacelle	Contains an ESP32 chip to manage the cargo release mechanism, receiving commands via ROS and converting them into serial signals for execution.
	VRPN Position Data	Shared between the motion capture, router, and DOC.
	Deliver Control Signals	Direct commands for the cargo release and receiving processes.
	Drone Status	Monitored and updated through the DOC and GPC.

(continued)

Table 1. *(continued)*

Drone	PX4 Flight Controller	Manages the drone's flight dynamics and stability by executing the Mavlink commands received through the Mavros middleware.
	Onboard Computer (DOC)	Acts as the central processing unit for the drone, handling ROS (Robot Operating System) communication. It subscribes to VRPN topics to acquire precise positional data and processes commands from the control computer.
	Nacelle	Contains an ESP32 chip to manage the cargo release mechanism, receiving commands via ROS and converting them into serial signals for execution.
	Received Signal	Confirmation of cargo reception sent back to the GPC.
	Remote Control Signal	Allows for manual intervention when required.
	Hardware Location:	Maintains the spatial awareness of all system components.

3 Workflow and Algorithms

3.1 Workflow Description

After launch the system, the GPC transmits a takeoff order to the drone via Simulink. Once the command is received, the drone will automatically unlock and take off from the appointed location within t0 seconds. After taking off, the drone ascends to a predetermined height of h0 meters and hovers at the origin of the spatial coordinate system of the motion capture. After t0 seconds, the drone commences its flight towards the target delivery point set in the algorithmic program.

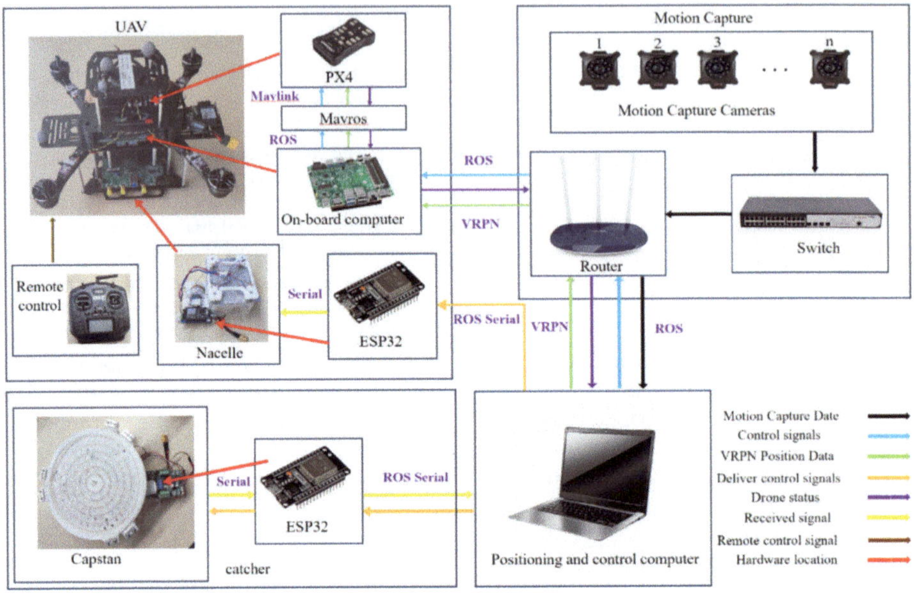

Fig. 1. System Components of the Drone Cargo Delivering System

During the flight, the drone constantly compares its real-time coordinates with those of the target point. If the horizontal position error is within ϵ meters and remains within this error range for t_1 seconds, without any alteration of the drone's attitude, it is determined that the drone has reached the target delivery point. Subsequently, the drone reduces its hovering altitude to h_1 meters above the ground and maintains its hovering status.

The program automatically governs the delivery device to release the cargo, discharging it from the nacelle. If the winch detects that the cargo has fallen within the detection area, it will automatically grasp the cargo. Once the cargo has been successfully grasped, the winch sends a signal to the DPC, indicating that the drone can return. Upon receiving the return signal, the drone will autonomously fly back to the original takeoff point and land safely, thereby completing the entire process.

3.2 Algorithm Descriptions

To effectively manage the position and attitude of the drone during its flight, the employment of a stable control system is of vital significance (Fig. 2). A quadcopter is inherently an unstable system, thereby highlighting the significant necessity of a control system. The current control framework of the controller operates in the following manner: A feedforward control command is generated by computing an angle or angular velocity control command and a throttle control command from the higher-order derivatives of the desired values within the transmitted control commands. Additionally, the desired values of the lower-order derivatives in the transmitted control quantities are utilized for cascaded feedback compensation. The system employs two controllers to handle position and two control loops for attitude. Specifically, there exists an outer-loop controller for position control (x, y, z) and an inner-loop controller for attitude control (ϕ, θ, φ).

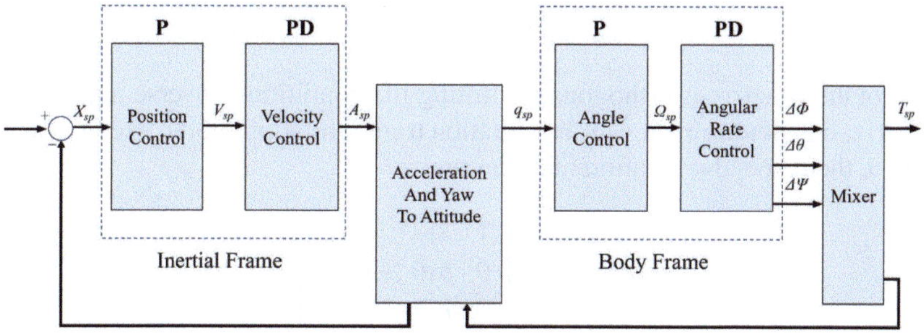

Fig. 2. Diagram block of proposed control system

In the ROS environment, however, only the external loop can be readily reprogrammed. As the external loop is accountable for position control, integral action is not required. This is because the existence of integral (reset) windup can result in considerable overshoot. Hence, a Proportional-Derivative (PD) controller is an appropriate choice for controlling the quadcopter. The control behavior of the PD controller can be expressed in (1),

$$un = K_p e_{x,y,z}(n) + K_d \Delta e_{x,y,z}(n). \tag{1}$$

3.3 Drone and Motion Capture Coordinate Transformation

For the motion capture [2], the position of any point p within the motion capture area can be represented by a 3 × 1 column vector $^A p$

$$^A p = \begin{bmatrix} p_x \\ p_y \\ p_z \end{bmatrix},$$

where p_x, p_y, and p_z are the three coordinate components of point p in the coordinate system [2], and $^A p$ is the position vector.

To describe the orientation of a rigid body B within the capture area, a 3 × 3 matrix $^A_B R$ consisting of the direction cosines of the three principal unit of coordinate system [6] relative to the reference coordinate system [2] is used. This matrix is known as the rotation matrix.

$$_B{}^A R = \begin{bmatrix} ^A x_B \, ^A y_B \, ^A z_B \end{bmatrix} = \begin{bmatrix} r_{11} \; r_{12} \; r_{13} \\ r_{21} \; r_{22} \; r_{23} \\ r_{31} \; r_{32} \; r_{33} \end{bmatrix}$$

The rotation matrix has nine elements, with the three column vectors ${}^{A}x_B$,${}^{A}y_B$ and ${}^{A}z_B$ being unit vectors that are mutually orthogonal, satisfying the orthogonality condition:

$${}^{A}x_B \cdot {}^{A}x_B = {}^{A}y_B \cdot {}^{A}y_B = {}^{A}z_B \cdot {}^{A}z_B = 1$$

$${}^{A}x_B \cdot {}^{A}y_B = {}^{A}y_B \cdot {}^{A}z_B = {}^{A}x_B \cdot {}^{A}z_B = 0$$

The rotation matrix is orthogonal, fulfilling the conditions: inverse matrix = transpose matrix, and determinant = 1. For rotation transformations about axes x, y, and z by an angle θ, the respective rotation matrices are:

$$R(x,\theta) = \begin{bmatrix} 1 & 0 & 0 \\ 0 & cos\theta & -sin\theta \\ 0 & sin\theta & -cos\theta \end{bmatrix},$$

$$R(y,\theta) = \begin{bmatrix} cos\theta & 0 & sin\theta \\ 0 & 1 & 0 \\ -sin\theta & 0 & cos\theta \end{bmatrix},$$

$$R(z,\theta) = \begin{bmatrix} cos\theta & -sin\theta & 0 \\ sin\theta & cos\theta & 0 \\ 0 & 0 & 1 \end{bmatrix},$$

If the coordinate system [6] has the same orientation as [2] but different origins, this is described by a translation coordinate transformation (Fig. 3). The displacement vector A_{pB0} describes its position relative to [2]. If the position of point p in coordinate system [6] is ${}^{B}p$, then its position vector relative to coordinate system [2] is given by vector addition:

$${}^{A}p = {}^{B}p + {}^{A}p_{B_0}$$

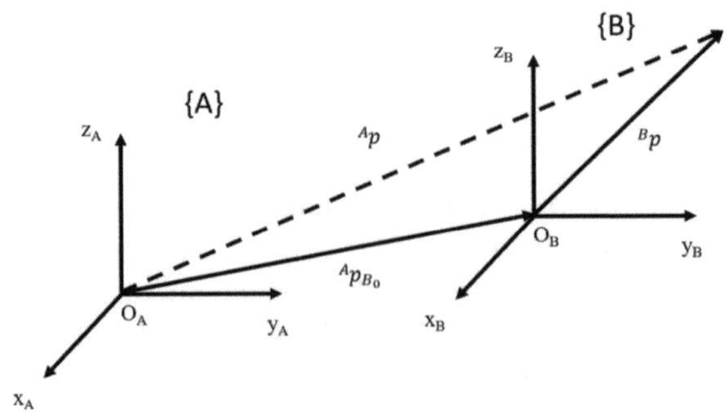

Fig. 3. Translation Transformation

If coordinate systems [6] and [2] have the same origin but different orientations, this is described by a rotation coordinate transformation (Fig. 4). The rotation matrix describes the orientation of [6] relative to [2]. The transformation relation for the same point p in both coordinate systems [2] and [6] is:

$$^{A}p = {}_{\mathrm{B}}{}^{\mathrm{A}}\mathrm{R}^{B}p$$

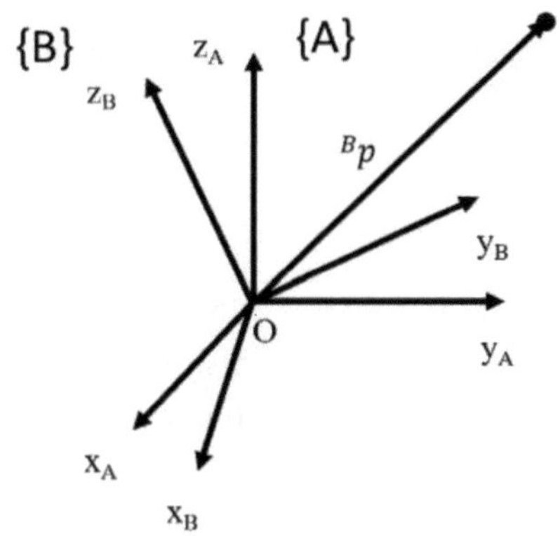

Fig. 4. Rotation Transformation

For the most general case where the origin of coordinate system [6] does not coincide with that of [2] and their orientations differ, the displacement vector $^{A}p_{B0}$ describes the origin of [6] relative to [2], and the rotation matrix $_{B}{}^{A}R$ describes the orientation of [6] relative to [2]. As shown in Fig. 5, the transformation relation for any point p in both coordinate systems [2] and [6] is:

$$^{A}p = {}_{B}{}^{A}R^{B}p + {}^{A}p_{B0}$$

This can be viewed as a composite transformation of coordinates.

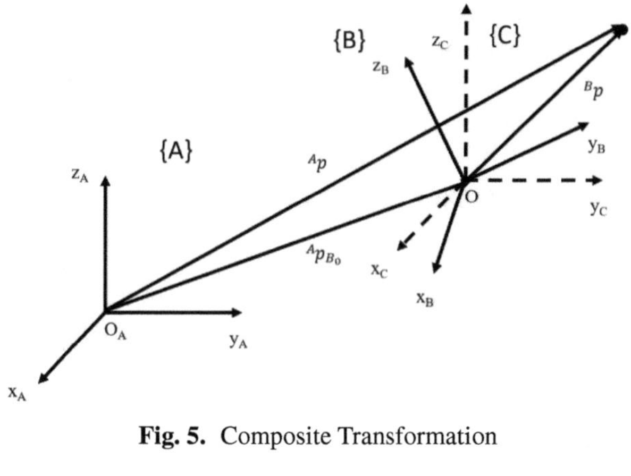

Fig. 5. Composite Transformation

4 Experimental Environment

The experiment was conducted in a controlled indoor environment to ensure precision and repeatability. The testing area was a motion capture zone measuring 25 m by 25 m, equipped with adequate lighting to maintain optimal testing conditions (Table 2).

Table 2. Parameters of hover height, duration and error

Parameters	Value
h0	1.5 m
h1	1 m
t0	15 s
t1	2 s
ϵ	0.1 m

As shown in Fig. 6(a) to (d), the drone program is initiated, and the takeoff command along with the winch's standby command for receiving the cargo is sent. The drone takes off, reaches an altitude of 1.5 m above the origin within 15 s, and hovers. Subsequently, it flies to the target point with coordinates (1, 1) in the xy−plane. During the flight, the system checks if the drone's horizontal position error relative to the target point is within 0.1 m and maintains this accuracy for 2 s. If the conditions are met, the drone is considered to have arrived at the target point. The drone then descends to an altitude of 1 m above the ground and hovers, sending the cargo release command.

Figure 6(e) to (h) illustrate the sequence following the receipt of the release command by the delivery mechanism. The electromagnet on the device is initially magnetized by default. The ESP32 controls the electromagnet via a solid-state relay. When the ESP32's control GPIO sends a high-level signal, the relay activates, powering the electromagnet off, which causes the magnetism to cease, thus releasing the cargo from the pod. When the winch below detects the cargo, the ESP32 drives the motor to rotate, securing the cargo to complete the receiving process. Simultaneously, it sends a message to the drone indicating that the cargo has been received and the drone can return. The drone then returns to the origin and lands.

5 Conclusion and Future Prospects

This study developed and tested a drone cargo delivering system, utilizing ROS framework for controlling. The results of multiple trials demonstrated the system's robust performance in indoor scenarios, where it reliably and successfully completed cargo hoisting tasks. The entire process, from takeoff to cargo delivery and return, was autonomously executed without the need for manual intervention, showcasing the system's capability for automated operation. The system exhibited high precision, reliability, and autonomy, efficiently performing cargo delivery tasks with small error margins, making it a promising solution for drone-based logistics in complex environments.

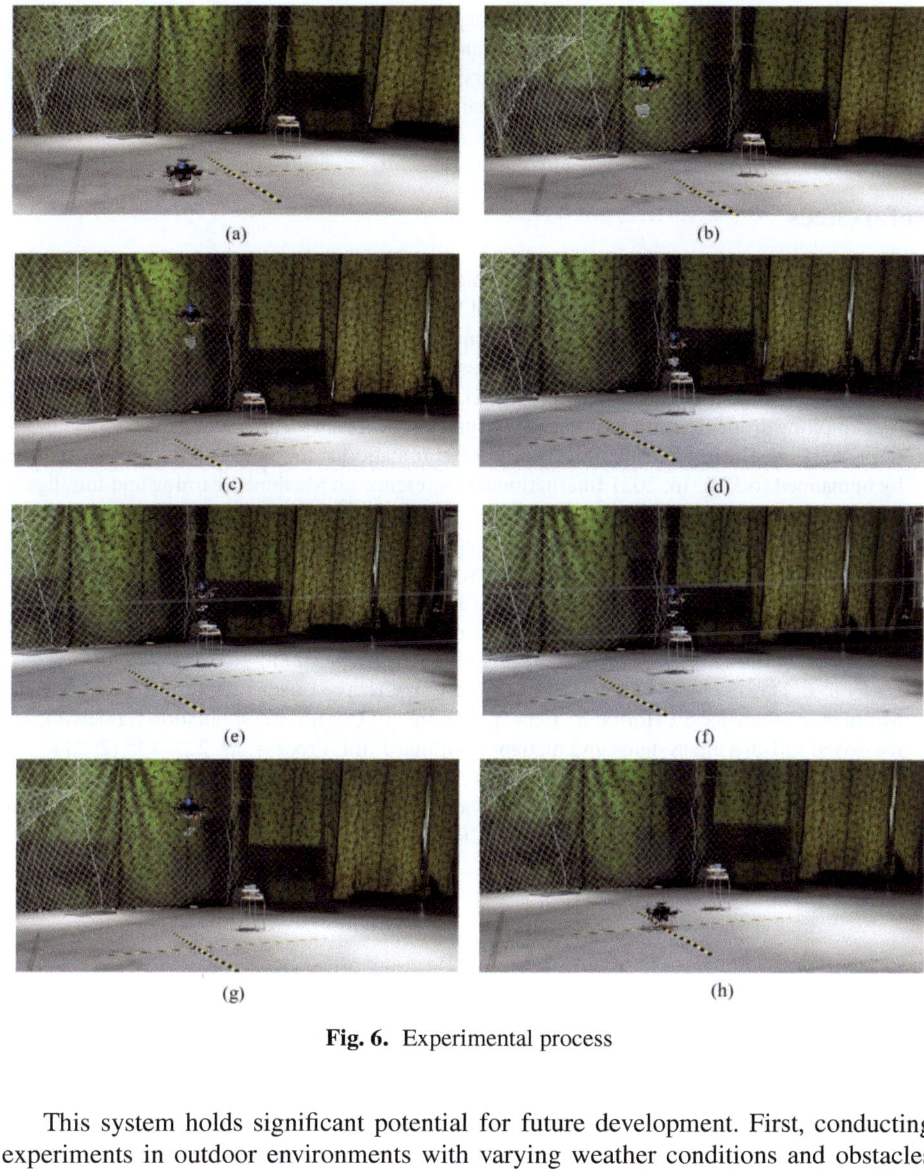

Fig. 6. Experimental process

This system holds significant potential for future development. First, conducting experiments in outdoor environments with varying weather conditions and obstacles will help assess the system's performance and robustness in real-world scenarios. Second, exploring machine learning techniques for adaptive control based on environmental feedback can enhance the control algorithms to improve efficiency and response time. Additionally, research into the system's scalability is essential to handle larger payloads and longer distances, as well as integration with multiple drones operating in a coordinated manner. Addressing these challenges will improve the system to meet the stringent requirements of commercial applications, paving the way for its widespread adoption in the logistics industry.

Acknowledgment. The work described in this paper was supported in part by the National Natural Science Foundation of China (NSFC) under Grants 62201363 and 62101335, in part by the Fundamental Research Funds for the Central Universities under Grants D5000220482, and in part by the Key Research and Development Plan of Shaanxi Province under Grant 2024GX-YBXM-301.

References

1. McCoy, J.V.: Unmanned aerial logistics vehicles: A concept worth pursuing?. Fort Leavenworth, KS: US Army Command and General Staff College (2003)
2. Leong, P.Y., Ahmad, N.S.: Exploring autonomous load-carrying mobile robots in indoor settings: a comprehensive review. IEEE Access (2024)
3. Deja, M., Siemiątkowski, M.S., Vosniakos, G.C., Maltezos, G.: Opportunities and challenges for exploiting drones in agile manufacturing systems. Procedia Manufactur. **51**, 527–534 (2020)
4. Cao, W., Dou, L.: Implementation method of automatic loading and unloading in container by unmanned forklift. In: 2021 International Conference on Machine Learning and Intelligent Systems Engineering (MLISE), pp. 503–509. IEEE (2021)
5. Li, C., Yu, H., Cao, S.: Indoor express delivery technology research and implementation: an application exploration based on quadrotors. In: International Conference on Autonomous Unmanned Systems, pp. 364–374. Springer (2023)
6. Barcelos, C.O., et al.: Robot formation performing a collaborative load transport and delivery task by using lifting electromagnets. Appl. Sci. **13**(2), 822 (2023)
7. Cannon, J., Downs, R.C., Orlando, A.: UAV to Carry Heavy Loads (2024)
8. Miura, K., Tokunaga, S., Horita, Y., Oda, Y., Azumi, T.: Cosam: Co-simulation framework for ros-based self-driving systems and matlab/Simulink. J. Inf. Process. **29**, 227–235 (2021)
9. Liu, C., Jiang, Z., Xu, R., Ji, X., Zhang, L., Qian, H.: Design and optimization of a magnetic catcher for UAV landing on disturbed aquatic surface platforms. In: 2022 International Conference on Robotics and Automation (ICRA), pp. 1162–1168. IEEE (2022)

New Power System Control and Power Equipment Design

Optimizing Li-Ion Battery Preconditioning Strategies for Electric Vehicles in Cold Climates Using Thin Film and Internal Heating Approaches

Adelina-Ioana Ilieş(✉) and Rajmond Jánó

Applied Electronics Department, Technical University of Cluj-Napoca, Cluj-Napoca, Romania
{adelina.ilies,rajmond.jano}@ael.utcluj.ro

Abstract. Cold temperatures significantly degrade the performance, efficiency, and lifespan of Li-ion batteries, posing a major challenge for electric vehicles (EVs) in cold climates. To mitigate these effects, effective preconditioning methods are necessary to heat the battery to an optimal operating temperature before use. Proper thermal management not only enhances battery reliability but also ensures consistent performance and longevity. This study investigates and compares two widely used preconditioning techniques for Lithium-ion batteries, external heating using thin film heaters and internal heating, through finite element method (FEM) simulations. By analyzing the effectiveness of both external and internal heating approaches, this research seeks to determine the most efficient method for reducing performance losses in EVs under cold-weather conditions. Additionally, a hybrid approach that combines both internal and external heating is investigated for its potential to minimize energy consumption while achieving rapid and uniform preconditioning.

Keywords: battery preconditioning · electric vehicle · low temperature

1 Introduction

Addressing climate change has become one of the most urgent global priorities, with the transportation sector being a major contributor to greenhouse gas emissions. Road transportation is a significant contributor to global CO_2 emissions, driving the automotive industry to adopt more sustainable solutions. In recent years, electric vehicles (EVs) have gained traction as an environmentally friendly alternative to traditional internal combustion engine vehicles. However, despite improvements in battery technology, power efficiency, and charging infrastructure, concerns about limited driving range, particularly in unfavorable weather conditions, continue to slow the widespread adoption of EVs [1].

EVs depend solely on lithium-ion (Li-ion) batteries for both propulsion and auxiliary power needs, including cabin heating. While modern EVs are capable of covering distances of up to 500 km on a single charge under ideal conditions, their actual range

T. Zielinska (Ed.): PEAE 2025, LNNS 1679, pp. 129–138, 2026.
https://doi.org/10.1007/978-3-032-08221-3_12

can fluctuate significantly due to environmental factors. Cold temperatures, in particular, pose a substantial challenge, as they not only increase energy demand for heating but also directly impact the efficiency of Li-ion cells. Studies indicate that EV range can decrease by 30–40% in cold weather, with extreme cases reporting reductions of up to 70% when temperatures drop to –26 °C or lower [2]. These performance losses are primarily due to the increased energy demand for thermal management and the reduced electrochemical efficiency of Li-ion batteries at low temperatures.

At subzero temperatures, the electrochemical reactions within Li-ion cells slow down, leading to higher internal resistance and reduced ionic conductivity. As a result, the battery's ability to store and deliver energy diminishes, causing a significant drop in both available capacity and power output. Severe cold conditions (–40 °C) can cause energy retention to drop by up to 95%, with power capability losses exceeding 98% compared to operation at room temperature [3]. Additionally, the sluggish charge transfer kinetics at low temperatures can lead to lithium plating, an undesirable process where metallic lithium accumulates on the anode instead of integrating properly into the electrode. This not only accelerates battery aging but also increases the risk of permanent capacity loss and potential safety hazards. To prevent damage, battery management systems (BMS) impose strict limitations on charging rates in cold conditions, extending charging times significantly and reducing the effectiveness of regenerative braking [4].

To address these challenges, effective battery thermal management is crucial for maintaining reliable EV performance in cold climates. One key strategy is battery preconditioning, which involves heating the battery to an optimal operating temperature before use. This can be achieved through various methods, including external heating elements such as resistive heaters, thin-film heating layers, and liquid-based thermal management systems that circulate heated coolant. Internal self-heating mechanisms, such as controlled micro-discharging and recharging, as well as heat generated from onboard power electronics, can also contribute to warming the battery. Additionally, waste heat from the electric motor can be redirected to assist in preheating. These approaches improve charge transfer efficiency, reduce internal resistance, and help prevent issues like lithium plating, ultimately enhancing battery performance and extending the driving range in cold conditions [5,6].

This study explores and compares two of the most popular preheating techniques for Li-ion batteries through finite element method (FEM) simulations. By evaluating the effectiveness of both external and internal heating strategies, this research aims to identify the most efficient method for minimizing cold-weather performance losses in EVs.

Among the external heating methods, thin-film heating has gained attention due to its lightweight, flexible design and ability to provide uniform heating. This technique utilizes thin resistive films placed around the battery, which generate heat when an electric current passes through them. Thin-film heaters can rapidly warm up battery cells with high energy efficiency and minimal weight addition. However, challenges such as potential thermal runaway risks and additional power requirements must be carefully managed to ensure safety and longevity [7, 8].

Internal preheating methods, on the other hand, rely on the battery itself to generate heat through its internal resistances, via controlled discharge/charge processes. Common strategies include constant current discharge (CCD) and constant voltage discharge (CVD), where the battery consumes a portion of its stored energy to warm up before operation. Higher discharge currents in CCD or lower discharge voltages in CVD can accelerate heating, but excessive values risk over-discharge and capacity degradation. Optimized discharge profiles, such as dynamic programming approaches, can improve efficiency by balancing heating time, energy consumption, and battery longevity. Internal heating is advantageous due to its low cost, simplicity, and reliability, as it does not require additional heating components. However, its efficiency remains relatively low, and it significantly depletes battery capacity during warm-up, making it viable mainly when the initial state of charge (SOC) is sufficiently high, or the EV is already connected to a charger (i.e.: during overnight charging at home) [9, 10].

The insights gained from this work contribute to the advancement of battery thermal management systems, supporting the broader goal of making electric mobility more viable and reliable across diverse climatic conditions.

2 Methodology

2.1 Proposed Goal

The goal of the research conducted is twofold. First, we would like to see the efficiency of the two preconditioning methods studied. Specifically, we are analyzing how long it takes to precondition an 18650 Li-Ion battery cell from 0 °C to 20 °C under different conditions, such as different power fed to the heater elements or different charging current used for the cell. Secondly, we are comparing the two preconditioning mechanisms studied: heating the cell using thin-film heating elements vs. self-heating generated by internal joule heating during charging/discharging of the batteries.

The final goal of the paper is to suggest the optimal strategy to be used in an EV for preconditioning the batteries before the charging/discharging cycle begins in order to optimize the charging speed as well as prolong the useful life of the battery cell itself.

2.2 Simulation Models

Firstly, the simulation model for an industry standard 18650 Li-Ion battery cell was constructed, as presented in Fig. 1. This was modeled by using a simple cylindrical shape with the material characteristics defined as presented in Table 1. These material properties for a standard 18650 Li-Ion cell were determined and validated in an earlier study [11].

Secondly, the thin-film heaters were added to the battery cells. These have a length of 60 mm and a thickness of 1 mm. Two heaters were attached centered to the Li-Ion cell model as seen in Fig. 2. The material for the heaters was set to copper with a Kapton (Polyimide) layer as the contact surface between the heaters and the battery casing. The thickness of the contact layer was modeled to be 0.1 mm with a thermal conductivity of 1 W/mK.

Table 1. Table captions should be placed above the tables.

Property [Unit of Measurement]	Value
Thermal Conductivity [W/mK]	3.35
Density [kg/m^3]	2725
Specific heat [J/kgK]	960

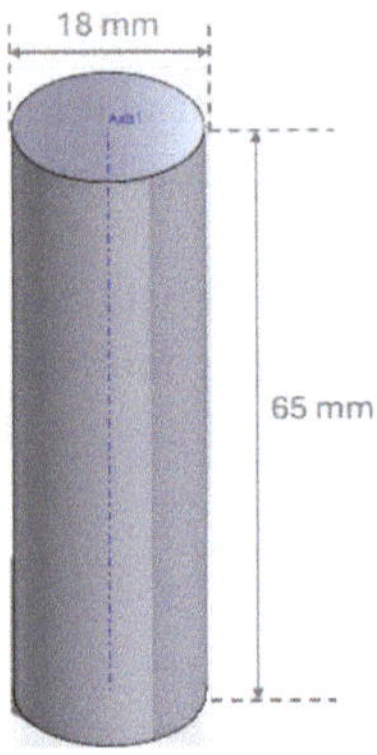

Fig. 1. Standard 18650 Li-Ion cel model

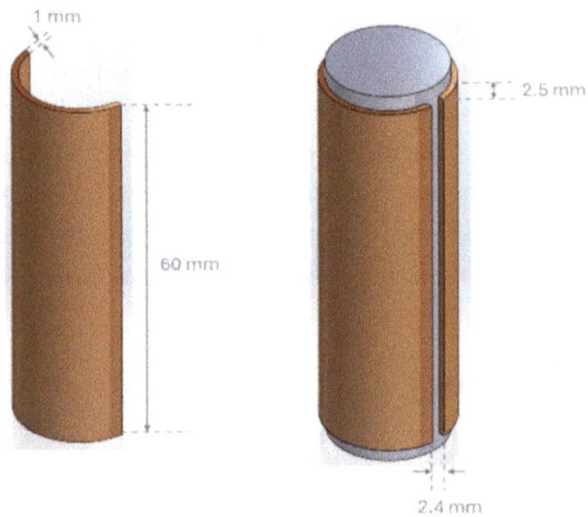

Fig. 2. Li-Ion battery cell model with thin-film heaters attached

For simulating internal heating no additional elements were added to the model, rather a volumetric heat generation rate was set for the body of the Li-Ion cell.

An additional consideration was that simulating the cell as presented in Fig. 2, with just the heaters around it, would mean that the simulator would place an infinite amount of air around the cell. This would be an unrealistic scenario: in real-world use cases the battery cells would always be installed inside of an electric vehicle, in some sort of case. While these casings would certainly not be completely airtight, the heating elements would still not need to heat infinite amounts of air around the casing. Simulating the battery preconditioning with an infinite amount of air around it would greatly reduce the efficiency of the preconditioning heaters and would not provide accurate enough results. Therefore, a non-airtight container was designed, in which the battery cell model was placed during simulations, as presented in Fig. 3. Since currently for our goals, we do not want to analyze the influence of this external casing, the material for it has been set to a perfect insulator. Small vents have been added to the bottom of the casing in order to simulate that the case is not airtight and to prevent pressure build-up inside the casing, which would happen when the air around the battery cell is heated and would influence the results and measured preconditioning times.

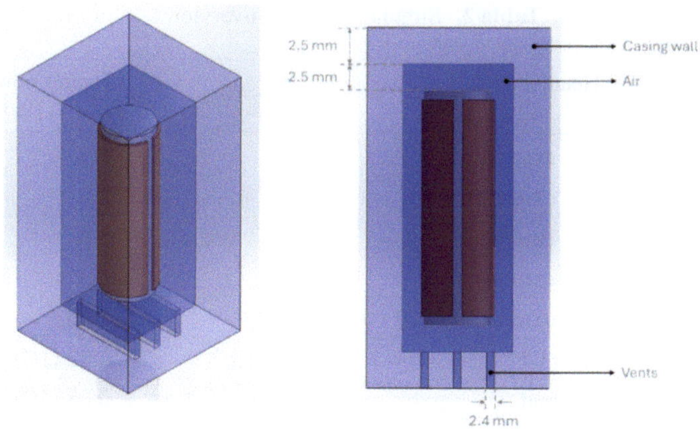

Fig. 3. Insulator casing added around the battery model

2.3 Simulated Setups

In order to evaluate the efficiency of the different preconditioning strategies FEM simulations using Cadence Celsius EC Solver were ran. The mesh was set to "Automatic" with an initial mesh level of 3. All solid materials as well as the surrounding air were set to an initial temperature of 0 °C. Then the different preconditioning strategies were activated and the time if took for the Li-Ion battery cell body to reach 20 °C was monitored. At this temperature the battery cell can be considered to be inside the optimal charge/discharge window where charging times are the fastest and cause the least degradation to the lifetime of the cell.

During the activation of the preconditioning heating mechanism the temperature of the surrounding air was not changed, and it was maintained at 0 °C. This would realistically simulate an electric vehicle, parked in a cold climate environment during the night, with a preprogrammed charging cycle that would start automatically at some point during the night.

The initial step was to evaluate the efficiency of the thin-film heaters when different power settings are applied to them. Therefore, the initial five simulations (Run #1 - #5) consisted of 0.5W, 1W, 2W, 5W, 10W of total power being applied to the thin-film heaters. After that, the effect of only internal heating was determined. For this the charging of the cell was simulated with a charging rate of 1C (2.2A), 2C and 3C. The internal resistance of the cell was set to 25 mΩ, an average value for 18650 Li-Ion cells. Finally, combined strategies were evaluated. The first combined strategy used both the external heaters at 0.5W total dissipated power as well as internal heating for charging at 2C, at the same time. Then, just the two external heaters were turned on at 0.5W for 1100 s, followed by keeping the cell warm using just internal heating (at 2C). Finally, the two external heaters were kept on at 0.5W for 700 s, followed by internal heating at 2C to keep the battery preconditioned. An overview of all the ran simulations is presented in Table 2.

Table 2. Simulation setups overview

Simulation Run	Heaters used	Heater power (total) [W]
#01	2 × external	0.5
#02	2 × external	1.0
#03	2 × external	2.0
#04	2 × external	5.0
#05	2 × external	10.0
#06	Internal	0.12 (1C)
#07	Internal	0.48 (2C)
#08	Internal	1.08 (3C)
#09	2 × external + internal (always ON)	0.5 + 0.48
#10	2 × external (ON: first 1100 s) + internal (ON after 1100 s)	0.5 + 0.48
#11	2 × external (ON: first 700 s) + internal (ON after 700 s)	0.5 + 0.48

3 Results

The overall results of the simulations ran are presented in Table 3. In this table the time needed for the cell to reach 20 °C (from 0 °C) is presented in seconds and then also converted in minutes for a better overview.

As can be seen from the results, the two external heaters are very efficient in preconditioning the cell, taking just 18 min when powered at 0.5W total, and a mere 1 min when using 10W of total power. However, while it is true that the goal is to have the battery cell preconditioned within the optimal charging/discharging temperature window, raising the temperature of the cell too fast might lead to thermal stress of the cell, which may have exactly the opposite effect which was desired and actually shorten the lifetime of the cell. Therefore, we need to make sure that the temperature rising slope (seen in Fig. 4) is gentle enough to not thermally stress the cell. As a result, the most gradual heating at 0.5W total heater power was used for the follow-up combined preconditioning strategies.

In the next three simulations the effects of internal heating were analyzed. In the first scenario, where the cell is charged at 1C (2.2A) the cell does not reach the targeted 20 °C, as thermal equilibrium sets in at 19.76 °C. To be noted that this temperature is still well inside the optimum charge/discharge temperature window and therefore could still be used for preconditioning. This scenario also has the gentlest temperature rising curve out of the three simulated, as can be observed in Fig. 5. When charging the battery at 2C the target temperature is reached in 38 min, while at a charging rate of 3C the same temperature is achieved in 18 min.

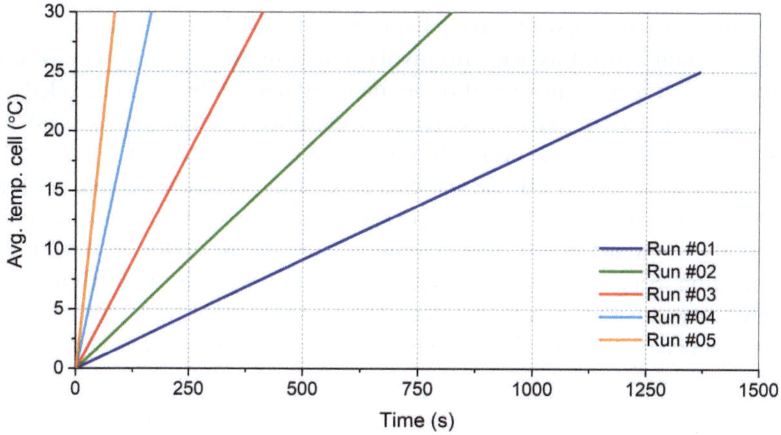

Fig. 4. Battery temperature cell evolution for runs #01 - #05 using only external heaters

Therefore, it can be deduced that using only self-heating for preconditioning is viable, the target 20 °C being reached (or almost reached) in most scenarios. However, since self-heating involves the heat dissipated by the cell while it is actively being charged, this method alone cannot be used for preconditioning, since preconditioning entails bringing the temperature of the cell within the ideal working temperature window before commencing the charging/discharging cycle. Nevertheless, this strategy can be used to maintain the cell temperature: once the battery has been preconditioned by active heating elements, self-heating should be sufficient in order to maintain the temperature of the battery inside the optimal working gap.

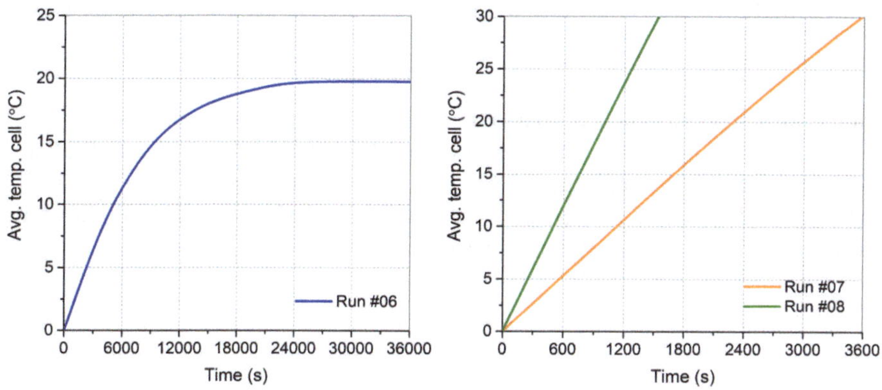

Fig. 5. Battery temperature cell evolution for runs #06 - #08 using only internal heating

In order to investigate how external and internal heating can efficiently complement each other, the next three simulations were analyzed. First, both the external heaters and internal heating were started at the initial point. The external heaters were set to dissipate 0.5W in total, while internal heating of the battery cell was set to be caused by charging at a 2C rate (0.48W). In this scenario the battery cell took 12 min to reach the desired 20 °C. However, again, since this scenario involves starting the charging of the cell while it is still too cold, it is not a preconditioning method that can be recommended.

In the second scenario, only the external heaters were turned on (at 0.5W) for the initial 1100 s. During this time period the cell reaches the aimed 20 °C. After this, the external heaters were turned off and the charging of the cell at a 2C rate commenced. As can be seen from the results presented in Fig. 6, the temperature of the cell continues to rise well above the aimed 20 °C, reaching a value of 38 °C after 1 h from the start of the simulation. Since this temperature might be too high for the Li-Ion cell, either the charging current is lowered, or an alternate strategy is employed, where the external heaters are stopped earlier, and the preconditioning is done partially by internal heating.

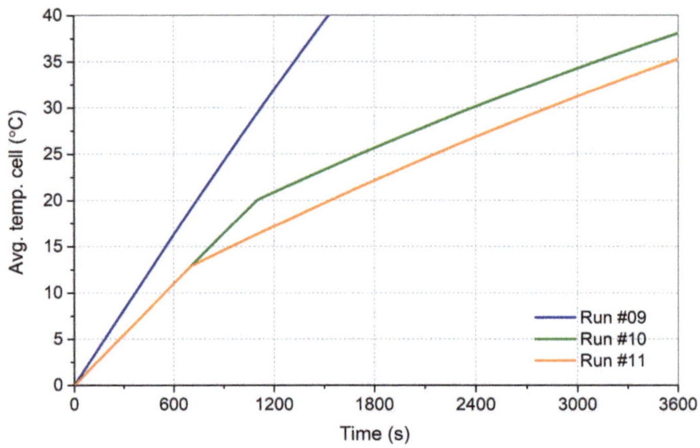

Fig. 6. Battery temperature cell evolution for runs #09 - #11 using combined preconditioning strategies

In the final scenario, the external heaters were active for only the first 700 s, after which preconditioning was continued by internal heating due to charging the cell at 2C rate. In this scenario, the cell reached 13 °C while actively heated before the charging commenced. After that, the temperature reached was 35 °C after 1 h from the start of the simulation, 3 °C less than the previous case.

Table 3. Simulation setups overview

Simulation Run	Heaters used	Cell preconditioning time (seconds)	Cell preconditioning time (minutes)
#01	2 × external	1092.5	18
#02	2 × external	547.5	9
#03	2 × external	274.5	5
#04	2 × external	111.0	2
#05	2 × external	56.5	1
#06	Internal	Not reached (thermal equilibrium at 19.76 °C)	
#07	Internal	2285	38
#08	Internal	1080	18
#09	2 × external + internal (always ON)	740	12
#10	2 × external (ON: first 1100 s) + internal (ON after 1100 s)	1100	18
#11	2 × external (ON: first 700 s) + internal (ON after 700 s)	1539	26

4 Conclusions

The purpose of the research presented in this paper was to evaluate the effectiveness of two preconditioning methods for 18650 Li-Ion battery cells: active preconditioning using thin film heaters and passive preconditioning using the Joule-effect self-heating of the cells while charging.

Using FEM Flow Simulations, it has been shown that using thin-film heaters is a very effective method for preconditioning the battery cells. Even with very low power consumption (0.5W), the cell can be brought inside the optimal working temperature window in around 18 min. This would offer an ideal scenario for preconditioning the cell at a slow enough rate to avoid thermal stress and the degradation associated with it.

It has also been demonstrated that using internal heating from the Joule-effect that occurs during the charging of the cell can also lead to the temperature of the battery rising to the ideal working window. However, the two disadvantages of this strategy are that (1) it is comparatively slow compared to using external heaters but also (2) can be used only when charging the battery, which is not ideal, as this entails starting the charging of the cell with it being outside of its ideal working temperature window.

Finally, we can conclude that in order to precondition the battery and maintain it within the optimal charge/discharge window, active heating should be used before charging is commenced. For a standard 18650 Li-Ion cell thin-film heaters can be used dissipating 0.5W for around 15 min, however the power can be increased for faster preconditioning times, if desired. After this period the active heating can be halted, the Joule-effect heating that occurs within the cell while charging being enough to maintain the temperature of the cell even in cold climates and freezing conditions. However, since internal heating can, in some cases, lead to the battery reaching higher than ideal temperatures, its state should be carefully monitored and the charging current adjusted by the battery management system.

Acknowledgement. This work was supported by the "Optimizing preconditioning methods to maximize Li-Ion battery performance" grant funded by the National Grant Competition - GNaC ARUT 2023.

References

1. Ji, Y., Wang, C.Y.: Heating strategies for Li-ion batteries operated from subzero temperatures. Electrochim. Acta **107**, 664–674 (2013)
2. Steinstraeter, M., Heinrich, T., Lienkamp, M.: effect of low temperature on electric vehicle range. World Electr. Veh. J. **12**(3) (2021)
3. Hu, X., Zheng, Y., Howey, D.A., Perez, H., Foley, A., Pecht, M.: Battery warm-up methodologies at subzero temperatures for automotive applications: recent advances and perspectives. Progress Energy Combust. Sci. **77** (2020)
4. Senol, M., Bayram, I.S., Naderi, Y., Galloway, S.: Electric vehicles under low temperatures: a review on battery performance, charging needs, and power grid impacts. IEEE Access **11**, 39879–39912 (2023)
5. Peng, X., Chen, S., Garg, A., Bao, N., Panda, B.: A review of the estimation and heating methods for lithium-ion batteries pack at the cold environment. Energy Sci. Eng. **7**(3) (2019)
6. Zhang, G., Ge, S., Xu, T., et al.: Rapid self-heating and internal temperature sensing of lithium-ion batteries at low temperatures. Electrochim. Acta **218**, 149–155 (2016)
7. Yang, W., Zhou, F., Liu, Y., Chen, X.: Preheating performance by heating film or the safe application of cylindrical lithium-ion battery at low temperature. Fire Technol. **59**, 1115–1135 (2023)
8. Lei, Z., Zhang, C., Li, J., Fan, G., Lin, Z.: Preheating method of lithium-ion batteries in an electric vehicle. J. Mod. Power Syst. Clean Energy **3**(2), 289–296 (2015). https://doi.org/10.1007/s40565-015-0115-1
9. Qu, Z.G., Jiang, Z.Y., Wang, Q.: Experimental study on pulse self–heating of lithium–ion battery at low temperature. Int. J. Heat Mass Transf. **135**, 696–705 (2019)
10. Jiang, J., Ruan, H., Sun, B., Wang, L., Gao, W., Zhang, W.: A low-temperature internal heating strategy without lifetime reduction for large-size automotive lithium-ion battery pack. Appl. Energy **230**, 257–266 (2018)
11. Jano, R., Ilies, A.I., Fodor, A.: Thermal simulations for 18650 Li-Ion batteries. In: 2022 IEEE 28th International Symposium for Design and Technology in Electronic Packaging (SIITME) (2022)

An IoT-Based Model for Detecting Line Hooking for Less Advanced Electricity Infrastructure: Preliminary Results

P. M. Kgaphola[1(✉)], R. T Hans[2], S. M. Marebane[1], and K. Sigama[3]

[1] Faculty of Information and Communication Technology, Tshwane University of Technology, eMalahleni 1035, Tshwane, Republic of South Africa
{kgapholapm,marebanesm}@tut.ac.za

[2] Computer Science Department, Tshwane University of Technology, Soshanguve 0152, Republic of South Africa
hansr@tut.ac.za

[3] Faculty of Information and Communication Technology, Tshwane University of Technology, Polokwane 0700, Republic of South Africa
sigamak@tut.ac.za

Abstract. Electricity theft, particularly through line hooking, poses a significant challenge in regions with less advanced electricity infrastructure. Traditional detection methods are often ineffective due to outdated infrastructure and manual enforcement approaches. This study proposes an IoT-based model integrating proximity sensors, object detection algorithms, and edge computing to enhance real-time detection and reporting of line hooking. Using a three-phase design incorporating Agent-Based Modelling, Discrete Event Simulation, and Tiny Machine Learning, the model ensures rapid detection, reduces false alarms, and enables autonomous theft prevention. Preliminary results demonstrate the model's potential to improve electricity theft detection, with future work focusing on implementation and evaluation in real-world scenarios. The proposed solution minimizes reliance on stable network connectivity by processing data locally using edge computing. This approach enhances the model's resilience and ensures continuous operation in areas with unreliable communication infrastructure. By addressing a critical gap in existing solutions, this research contributes to more effective electricity theft prevention strategies tailored for developing countries.

Keywords: Electricity theft detection · IoT and Edge Computing · line hooking

1 Introduction

Electricity theft has emerged as a significant challenge in the world, more so in developing countries where the electricity infrastructure remains unimproved [1]. This issue is worsened by the prevalence of line hooking, a method where individuals illegally connect to power lines, bypassing the formal electricity distribution system [2]. The consequences of such theft lead to substantial financial losses for utility companies and

T. Zielinska (Ed.): PEAE 2025, LNNS 1679, pp. 139–152, 2026.
https://doi.org/10.1007/978-3-032-08221-3_13

municipalities, increased operational costs, and a strain on the unmodernized infrastructure relied upon for electricity distribution [3, 4]. The current infrastructure in developing countries characterized by its age and inefficiency, is poor to handle the technological advancements of the modern era, making it increasingly vulnerable to theft and misuse [5]. Traditional electromechanical meters, for example, which are still commonly used in these areas, are unable to identify illegal connections or meter tampering in real time of which modern infrastructure can do. Furthermore, theft using techniques like line hooking frequently goes undetected for long periods of time, therefore, putting more strain on the grid [4].

Despite the severity of line hooking, existing solutions in the literature often require electricity consumption data from smart meters and advanced metering infrastructure (AMI) to detect theft [6]. As a result, they (current solutions) fail to address the unique challenges posed by the infrastructure setup in developing countries. For example, the outdated infrastructure in developing countries cannot pinpoint areas where actual electricity consumption exceeds recorded electricity consumption [7]. The inability to track such activity highlights the urgent need for innovative approaches tailored to the specific conditions of the setup in developing countries. The integration of Internet of Things (IoT) technology presents a promising avenue for addressing electricity theft. By leveraging IoT solutions, it is possible to enhance monitoring capabilities, improve detection of unauthorized connections, and ultimately reduce the rate of line hooking [8].

Based on the preceding discussion, this study proposes an IoT-based model that integrates proximity sensors, object detection algorithms, and edge computing to effectively address line hooking in areas with less advanced infrastructure, which normally exist in developing countries, as indicated in the previous paragraph. This study's contribution to literature lies in the proposed model. To achieve its objective, this study seeks to answer the following research question:

How can a model for detecting line hooking be designed for implementation using IoT technology in regions with less advanced electricity infrastructure?

The remainder of the study is organized as follows: First, the literature review section provides a background of the study and references related work. Next, the methodology section outlines the study's approach and model design. This is followed by the preliminary results and discussion section, which presents the findings from one phase of the proposed model's multiphase design and interprets these results. Finally, the conclusion summarizes the study.

2 Literature Review

Electricity theft is a global challenge affecting both developed and developing countries, with varying degrees of severity and solutions. While developed nations have implemented advanced technologies like SG and AMI to address theft, many developing countries, including South Africa, Pakistan, and India rely on labour-intensive methods to detect theft due to the limitations of their unimproved infrastructure [9]. In these countries, where aging infrastructure and socio-economic challenges prevail, conventional measures such as physical inspections, disconnections, and public awareness campaigns have proven insufficient. Furthermore, efforts such as government policies and subsidized electricity also fail to address the root causes of electricity theft more especially line hooking [7].

As developing countries continue to grapple with the dual challenges of energy theft and infrastructural inadequacy, the implementation of IoT-based solutions could serve as a critical step towards revitalizing the electricity sector and ensuring equitable access to this vital resource [5]. Research shows that technology-driven approaches like IoT are more effective in detecting and preventing electricity theft [7, 10]. The IoT systems act autonomously eliminating the need for the use of ineffective conventional methods, such as, physical inspections, for preventing electricity theft. These technology-driven approaches are able to detect and report electricity theft and even cut off power supply when electricity theft is detected [11, 12]. Additionally, they employ edge computing to facilitate immediate theft detection.

Edge computing decentralizes data processing, bringing computation closer to the source. This reduces latency, enhances real-time capabilities, and alleviates reliance on unstable networks [13]. In healthcare IoT-based systems, edge computing has significantly enhanced response times and optimized resource utilization [14]. For example, in remote patient monitoring systems, edge computing processes data from wearable devices locally, reducing latency and enabling real-time alerts for critical health conditions such as abnormal heart rates or sudden fall [15]. This study agrees with Haq, Pei, Zhang, Jianjun and Ahmad [10] when they say that, in the context of electricity theft detection, edge computing ensures uninterrupted theft detection by ensuring that detection and mitigation happens immediately without being affected by network connectivity, which is often needed to connect to a remote server for the purpose of analytics and reporting. Therefore, the proposed model in this study will implement edge computing to overcome challenges with delayed detection and reporting of electricity theft by IoT systems that use cloud computing.

However, the existing solutions largely focus on meter tampering, which involves physically manipulating an electricity meter to alter or reduce the recorded consumption, and meter bypassing, which eliminates the meter's role in measuring electricity consumption [20]. Furthermore, the provided solutions in the literature often neglect line hooking, a dominant form of electricity theft in developing countries. Line hooking poses unique challenges, as it takes advantage of the lack of real-time detection mechanisms and is tailored to exploit specific infrastructural weaknesses [15]. While some models, such as those utilizing data analytics address this issue in controlled settings, their reliance on advanced infrastructure renders them impractical for the outdated grid in developing countries [2, 12].

2.1 Related Work

Recent studies highlight the potential of IoT technologies to provide targeted solutions for line hooking. For instance, Rajasekaran, Ajay, Bhanu and Swetha [18] developed an IoT-based model employing infrared sensors to detect objects in proximity to power lines and capture real-time evidence of theft. However, their model could not differentiate between humans and non-human objects such as birds or cats that can come near the distribution box without causing it harm, leading to false alarms. On the other hand, Engelbrecht, Hanck and Osifeko [19] classified fraudulent users among those who are legally connected to the grid using IoT and the Support Vector Machine (SVM) algorithm. They achieved high accuracy, but they neglected to account for electricity theft by those who are not legally connected to the grid.

Mohanty, Iqbal and Thampi [20] further explored IoT-based technologies by designing a system capable of monitoring discrepancies between current levels at the source and consumer ends. Their approach involved using current sensors and Global System for Mobile Communications (GSM) modules to report theft in real time. While effective in identifying irregularities in electricity use, the model depended on stable cellular network connectivity as it needed to communicate with a cloud server for theft analytics, a limitation in regions with unstable networks like South Africa. This study addresses this challenge by incorporating edge computing to ensure uninterrupted detection and response.

Similarly, another innovative approach was demonstrated by Kumarana, Ananthib, Saranyac, Priyadharshinid, Thiviyabalae and Vaishnavif [11], who developed an Arduino-based system for detecting electricity theft by monitoring voltage and current anomalies. Their model used relays to disconnect illegal connections and notify relevant authorities. Although their solution could detect illegal connections, it also relied on stable network connectivity for determining electricity theft.

Based on insights from the above studies, this study aims to address their shortfalls by delivering a comprehensive and practical IoT-based solution tailored to developing countries infrastructure setup. The solution is a three-phase design for an IoT model that will solve the identified limitations. Through real-time detection and prevention of line hooking, the proposed solution will contribute to reducing electricity theft and improving the reliability of the developing countries power distribution networks. The three-phase design is discussed in the next section.

3 Methodology

This study proposes a design for an IoT-based model to detect and report electricity theft, specifically focusing on illegal hooking in regions with less advanced infrastructure, using South Africa as a case study. The design incorporates simulation, machine learning, and edge computing technologies to develop an effective, reliable, and context-sensitive solution. The proposed model addresses electricity theft, which is carried out through line hooking. The design will be executed in three distinct phases, each focusing on a specific aspect of the proposed model. The distinct phases will be designed using different simulation tools and training a machine learning (ML) model because no single tool can adequately cover all the aspects of the proposed model which are vital for creating a robust line hooking model that addresses the shortfalls of the solutions mentioned in the previous section.

Figure 1 gives a view of the distribution pole, and a possible structure of the electronic components arrayed in a realistic way to show how the proposed model might look like once implemented. The figure was created using Adobe Photoshop to show the actual components that will be used during implementation of the envisaged line hooking detection model. However, it is worth noting that during implementation the positioning and structure of the electronic circuit will differ from the depiction in Fig. 1.

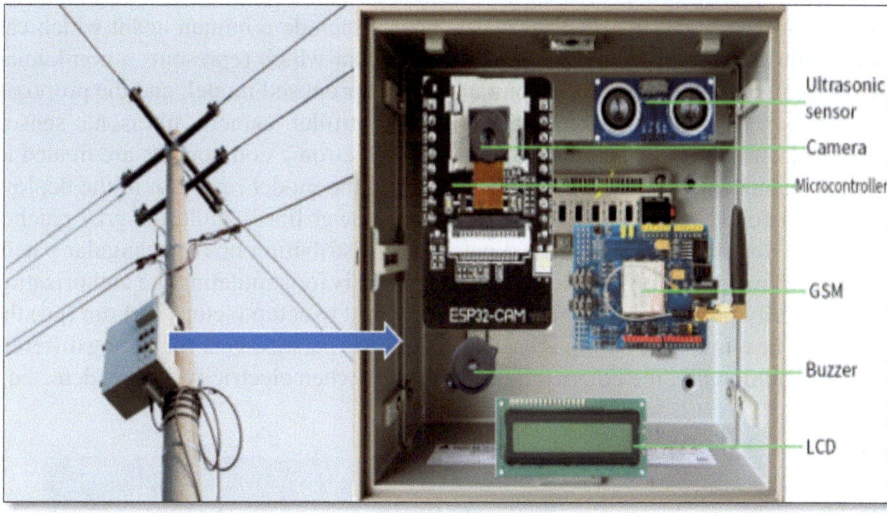

Fig. 1. Depiction of the proposed line hooking detection model.

The first aspect of the proposed model to be covered in the three-phase design is the environmental context in which the envisaged model will be deployed. This phase will show how the proposed model will respond to activities that will take place within the deployment environment of the proposed model. It will showcase how the proposed model will respond to theft attempts, non-human activity taking place within the detection range of the proposed model, and authorized personnel activity. The second phase of the design will give a more technical view of the detection mechanism, covering the functioning of the electronic components used in the proposed model, and how they communicate with each other to respond to activity that would have been detected by the proposed model's sensors. This phase gives a detailed view of how the proposed model responds to activity that occurs in phase1 in a systematic manner. Finally, the third phase will handle validation of theft attempts using ML. In this phase, object detection will be used to classify human, and non-human entities present within the detection range of the model. The image classification will be used to determine if there are actual theft attempts by humans or non-human entities that cannot harm the distribution system are being detected, therefore, accurately reporting electricity theft.

The following subsectins will futher expand on the the three-phase design for the proposed line hooking electricty theft detection model.

3.1 Phase 1: Agent-Based Modelling (ABM)

To cover the first aspect of the model, Phase 1 involves Agent-Based Modelling (ABM) to simulate interaction between various entities, including unauthorized intruders, authorized personnel, non-human entities, and the proposed model components. Using NetLogo software, the simulation is aimed at modelling the real-world deployment environment around electricity distribution poles. Figure 2 displays the NetLogo design of the ABM which comprises of agents, a switch, buttons, and patches resembling both the distribution pole and box in Fig. 1. The agents include a human agent which can be in an authorized or unauthorized state, a bird agent which represents a non-human object that can come within a detection range of the proposed model, and the proposed model component agents that include the microcontroller, camera, ultrasonic sensor, GSM module, and buzzer. The proposed model's electronic components are treated as individual agents and are responsible for triggering the model response in the deployment environmental. The black patches represent power lines, while the grey patches represent the distribution pole (vertical patch) and distribution box (rectangular patch) respectively. The 'detection-enabled' switch in Fig. 2 is for simulating the authorization process of utility personnel while the buttons are used to setup (setup) and run (go) the simulation. The climb pole button is used to control the person agent for testing different scenarios. Additionally, the command centre reports when electricity theft is detected.

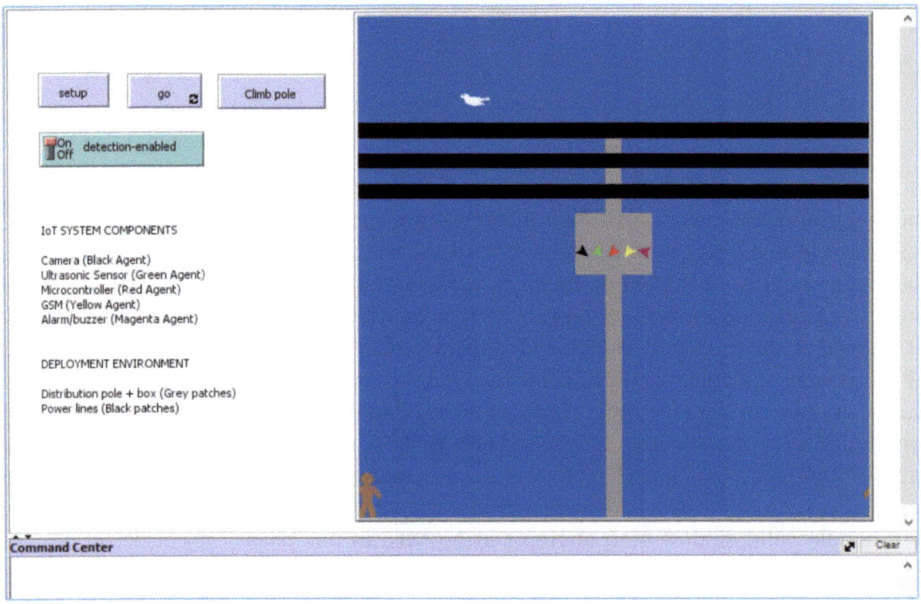

Fig. 2. ABM design for the proposed model.

3.2 Phase 2: Discrete Event Simulation (DES)

The second phase employs Discrete Event Simulation (DES) to model the sequential operation of the proposed model. This phase, to be conducted using Proteus software, focuses on simulating the behaviour of the electronic circuit's components involved in the proposed model. The circuit will be integrated into the distribution system so that it can be powered. Furthermore, the circuit will either be placed under the circuit brakers inside the distribution box or it will have its own casing situated under the distribution box. The positioning of the circuit will be subject to the municipality and utility companies' requirements which will be discussed in the implementation phase.

The goal of Phase 2 is to simulate the physical hardware circuit that connects and controls all key system components, including the ultrasonic sensor, ESP32 microcontroller, camera module, GSM SIM900 module, alarm system, LCD display, and backup power supply.

This simulation will help verify whether the system can correctly respond to specific real-world events. For example, when the ultrasonic sensor detects an object, it should trigger the ESP32 to activate the camera and capture an image of the scene. The captured image will then be processed locally, leading to a decision-making step where the system determines whether the detected object is a human (potential theft) or a non-human entity. If a human presence is identified, the system should automatically trigger the GSM module to send an alert message, activate the alarm system to deter the intruder, and update the event log for recordkeeping. Conversely, if the system determines the event was caused by a non-human object (such as an animal) or authorized personnel, no alert or alarm will be issued, ensuring proper filtering of false triggers and minimizing unnecessary responses.

3.3 Phase 3: Machine Learning Integration

The final phase integrates a Tiny Machine Learning (Tiny ML) model for real-time object detection. Tiny ML refers to the deployment of machine learning models on microcontrollers and other resource-constrained devices [21]. This technology enables intelligent data processing at the edge, allowing devices to operate with minimal power, memory, and computational resources [22]. Faster Objects, More Objects (FOMO), a lightweight object detection algorithm designed for resource-constrained devices, such as microcontrollers will be used in this study. Unlike traditional object detection models like You Only Look Once (YOLO) or Single shot detector (SSD), FOMO focuses on high-speed inference while maintaining the ability to detect multiple objects in a single image, making it the ideal tool for real time detection [21]. The researchers of this study plan to train and deploy a FOMO-based TinyML model using the GEINTRA Overhead ToF People Detection 3 (GOTPD3) dataset, incorporating 9,887 depth and infrared images captured under varied human activity conditions. Once the model is trained, we will conduct thorough evaluations of performance metrics, including accuracy, precision, recall, inference time, and microcontroller memory footprint. These tests will ensure the model's ability to reliably differentiate between human and non-human activity in real-time and to operate effectively on an ESP32-CAM microcontroller. Additionally, future work will include field trials in real-world deployment settings to assess the system's robustness under varying environmental conditions (such as weather, lighting, and sensor noise) and scalability to larger networks of distribution poles. These trials will provide valuable insights for refining the model and validating its performance beyond controlled simulation environments.

Through these efforts, we aim to advance the proposed model into a fully implemented and field-tested solution for electricity theft detection in regions with less advanced infrastructure.

By adopting the three-phase approach, the proposed design ensures a comprehensive and robust solution to electricity theft detection in South Africa and developing countries with similar infrastructure challenges.

3.4 System Design

The model design integrates several hardware and software components to achieve its objectives. Key components include an ESP32-CAM microcontroller, GSM SIM900 module, an ultrasonic sensor, a camera module, an alarm system, a liquid crystal display LCD screen, a Tiny ML model, and a backup power supply as illustrated in Fig. 3. The ultrasonic sensor will detect activity around the distribution box, triggering the camera to capture images of the area. The images will be processed by the FOMO model to validate theft attempts. If theft is confirmed (e.g. unauthorized human is detected), the model will send an SMS alert with the location and photo evidence to relevant stakeholders using GSM technology, activate an alarm to deter intruders, and log the theft incident. To address connectivity issues, the model will employ edge computing, enabling local data processing, namely, running object detection on the microcontroller to determine if actual theft is happening or non-human entities such as birds or cats are being detected. Classifying the captured images into theft attempts or non-theft attempts on the microcontroller instead of the cloud alleviates reliance on network connectivity for the model to respond to theft attempts.

Simulations will be conducted to replicate real-world operations and evaluate the model's effectiveness under different scenarios. These scenarios will include responses to authorized personnel, false detections and actual theft attempts. False detections can happen in rare cases where the ultrasonic sensor's receiver picks up acoustic sounds such as whistling or hissing sounds that travel at the same frequency as that of the ultrasonic sensor. Upon receiving the sound wave which the sensor will think is an object, the proposed model will trigger the camera to take a photo of the poles surrounding. In such a case, if there is a human at the base of the distribution pole the proposed model will likely classify the incident as a theft attempt because the ML model would have received an image of a human and determined that it is theft. The findings from these simulations will guide iterative improvements to the model's design, ensuring its robustness and reliability.

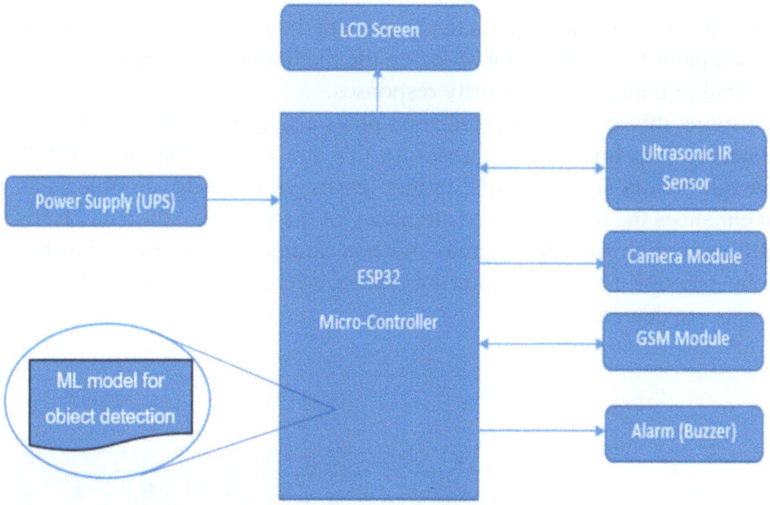

Fig. 3. Block diagram for the proposed model.

4 Preliminary Results and Discussion

This section presents the preliminary results from the implementation of the first phase of the three-phase design, conducted through an ABM approach. The ABM model was developed and tested to analyse how the proposed model would perform under various conditions within the intended deployment environment. These tests were essential for evaluating the model's effectiveness in detecting, categorizing, and responding to different types of interactions within the detection range from the distribution box.

To assess the model's functionality, a series of scenario-based tests were conducted, each designed to simulate real-world situations that the model might encounter. The tested scenarios include the detection of unauthorized access attempts to the distribution box, the handling of authorized personnel accessing the box, and the identification of non-human entities within the model's detection range. Each of these scenarios is critical for ensuring the model's reliability, accuracy, and ability to distinguish between authentic and potentially harmful interactions.

4.1 Detecting Non-Human Objects

The detection of non-human objects was tested by allowing a simulated bird agent to pass near the distribution box while the detection-enabled switch was activated, as illustrated in Fig. 4. The bird agent was programmed to move in various directions around the distribution box, including flying above, passing below, and crossing directly in front of it. These different positions were chosen to assess the model's ability to correctly distinguish between human and non-human entities within the detection range.

During the experiments, the model consistently interpreted the presence of the bird as a non-threatening event and appropriately avoided triggering any unauthorized access alerts or theft reports. The absence of a theft report in the command centre, as shown in

Fig. 4, confirms that the model effectively classified the bird agent as a nonthreatening entity. This capability is crucial for reducing the likelihood of false alarms, which could otherwise lead to unnecessary security responses.

Furthermore, this accurate classification plays a significant role in minimizing false alerts related to line hooking. By ensuring that the model does not mistakenly flag environmental factors such as birds, falling leaves, or wind-driven objects as security threats, it enhances the reliability and efficiency of the theft detection mechanisms. This refined differentiation between human and non-human interactions contributes to the overall robustness of the model, ensuring that security personnel are only alerted when actual threats are detected.

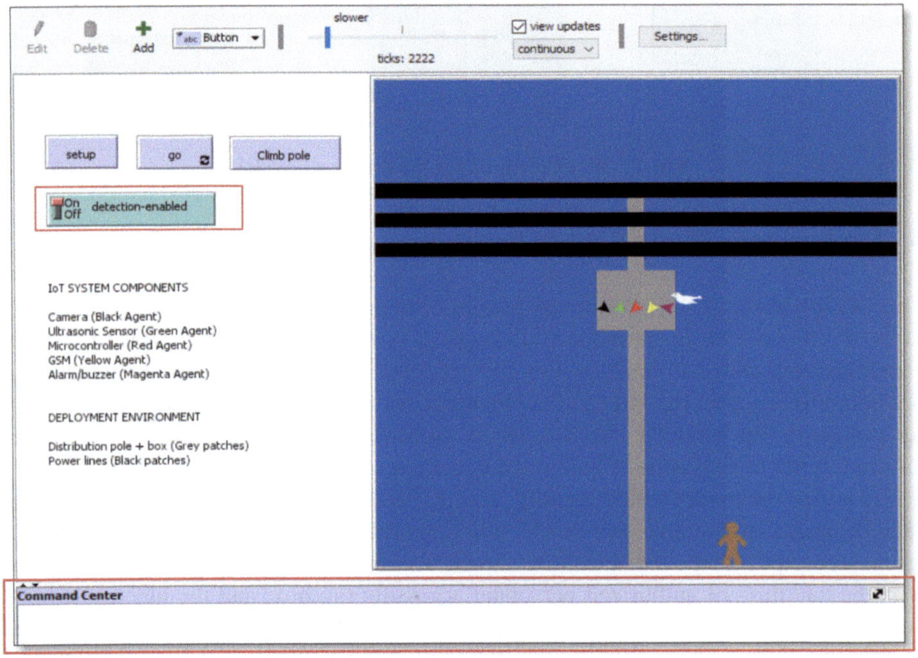

Fig. 4. Detection of non-human objects.

4.2 Detecting Unauthorized Access to the Distribution Box

To assess the model's effectiveness in accurately identifying theft attempts, a controlled test was conducted with a human agent interacting with the distribution pole while the detection-enabled switch remained activated. The test scenario involved triggering the human agent to climb up the distribution pole, entering the designated detection zone by using the climb pole button. The detection zone was dynamically measured using the built-in *in-radius* function in NetLogo, which calculates the relative proximity of the human agent to the sensor agent. As soon as the human agent entered the detection zone, a method called *process-image* was called which determines what type of agent was in the radius of the detection zone. The model successfully identified the activity as a potential theft attempt. It promptly flagged the incident, generated an alert, and reported it as theft, as illustrated in Fig. 5. This demonstrated the model's capability to detect unauthorized access in real-time, reinforcing its effectiveness in identifying security threats to the distribution infrastructure.

Moreover, in alignment with the model's design, no alert was triggered when the human agent merely passed underneath the distribution box without climbing. This was an intentional feature aimed at reducing false alarms in scenarios where an individual is present near the infrastructure but not engaging in any harmful activity. By ensuring that only interactions within the defined detection zone lead to alerts, the model enhances the reliability of theft detection.

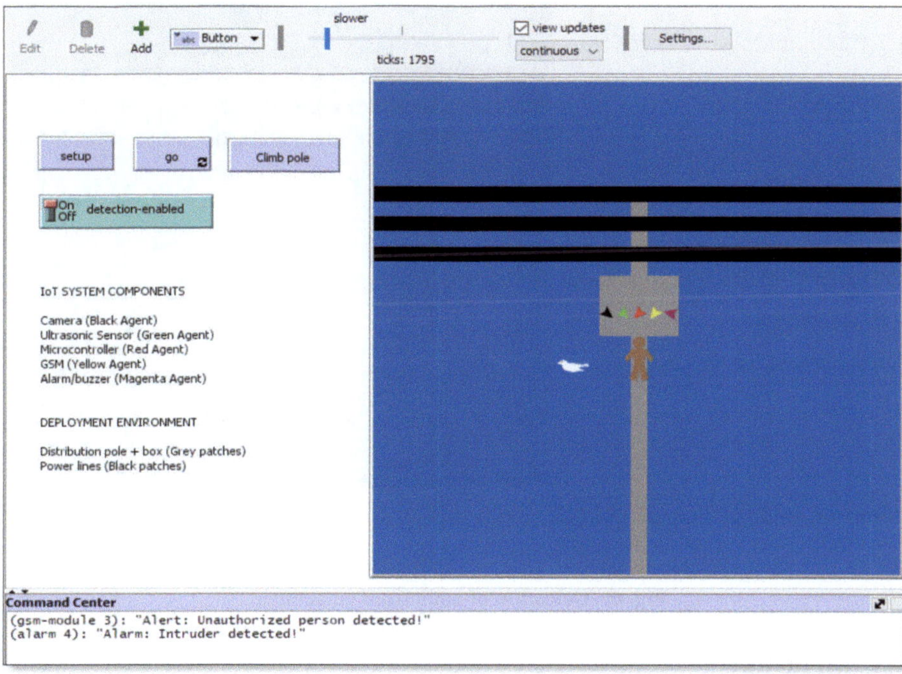

Fig. 5. Detection of electricity theft attempts

4.3 Detecting Authorized Access to the Distribution Box

On the other hand, to evaluate the model's ability to recognize authorized personnel, a test was conducted where the detection-enabled switch was intentionally turned off. With detection turned off, the model allowed the human agent to climb into and remain within the designated detection range without triggering an alarm. This confirmed the expected functionality when the detection feature is disabled, no alerts are generated, and no theft attempts are recorded, regardless of the human agent's proximity to the distribution box, as demonstrated in Fig. 6.

This test was crucial in verifying that the model operates flexibly based on security requirements, ensuring that authorized personnel can perform their tasks without unnecessary interruptions when detection is intentionally disabled. Once the detection mechanism was re-enabled, the model resumed its standard operation, immediately identifying the human agent within the detection zone. The model accurately triggered an alert and recorded the event as a theft attempt, reinforcing its ability to respond appropriately when security measures are active. This controlled evaluation highlights the model's adaptability and effectiveness in managing access under different conditions.

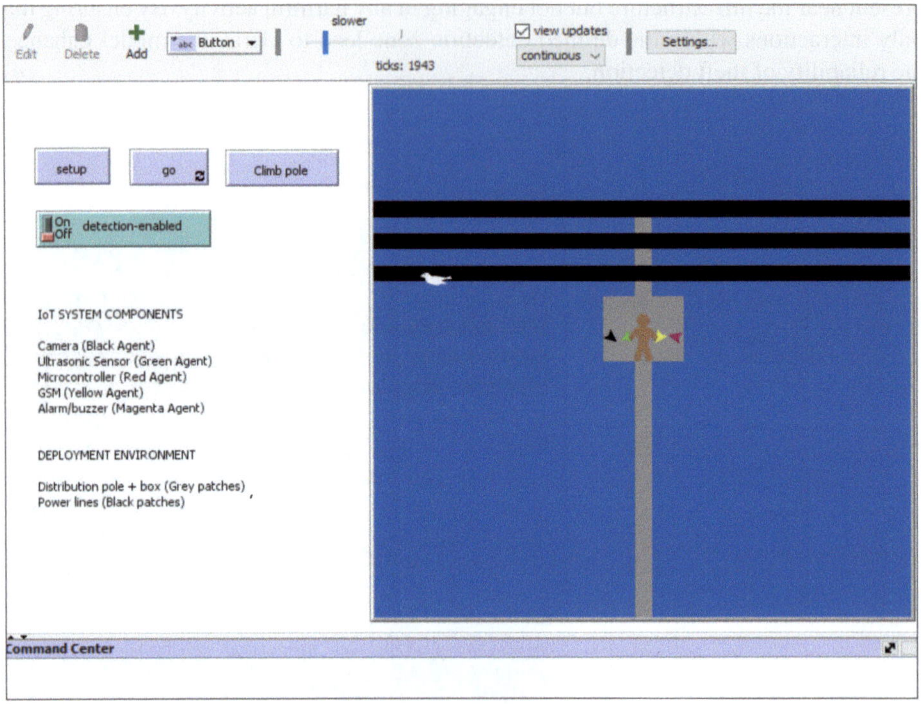

Fig. 6. Detection of authorized personnel.

5 Conclusion

This study presents a comprehensive design for an IoT-based solution tailored to the challenges of detecting electricity theft through line hooking in regions with limited grid infrastructure and network connectivity. By integrating Agent-Based Modelling, Discrete Event Simulation, and TinyML technology, this approach promises a real-time line hooking detection and prevention model with minimal false alarms. Preliminary work was done on the proposed solution by implementing one phase of the three-phase design through ABM. The implemented phase showcases how the proposed solution will behave in the environment that it will be deployed in. The Preliminary results show the model's ability to accurately distinguish between human and non-human entities, ensuring high accuracy and reliability in detecting theft attempts.

However, while the preliminary implementation of Phase 1 using ABM demonstrates promising results in accurately distinguishing between human and non-human entities and flagging potential theft attempts, several limitations remain. The simulation was conducted under controlled conditions that may not fully represent the complexity of real-world scenarios, where extreme weather conditions, sensor errors, and varying user behaviors can significantly influence the proposed models' performance. Moreover, the reliance on fixed detection zones and the assumption of stable sensor operation might lead to challenges when scaling the model to larger, more diverse deployment settings. Future work will need to address these limitations through extensive field trials, enhanced model tuning, and robust fault tolerance measures to ensure consistent reliability in different environments.

The design represents an evolving solution tailored to South Africa's grid infrastructure and lays the groundwork for future innovations in electricity theft prevention. Once fully implemented, this model has the potential to significantly reduce electricity theft, improve grid stability, and support equitable electricity distribution. Future work will focus on completing the remaining design phases, implementing the proposed model, and conducting field trials to assess its performance in real-world scenarios.

Disclosure of Interests. The authors declare no conflicts of interest.

References

1. Dube, F., Moyo, C.: The right to electricity in South Africa. Potchefstroom Electr. Law J. (PELJ) **25**, 1–21 (2022)
2. Zulu, C.L., Dzobo, O.: Real-time power theft monitoring and detection system with double connected data capture system. Electr. Eng. **105**(5), 3065–3083 (2023)
3. Mmako, M.M., Maribe, P.P.: The influence of infrastructure financing on rural municipalities in limpopo province, South Africa: issue for consideration. Int. J. Soc. Sci. Res. Rev. **7**(4), 366–377 (2024)
4. Shokoya, N.O., Raji, A.K.: 'Electricity theft: a reason to deploy smart grid in South Africa. In: Editor (Ed.)^(Eds.): 'Book Electricity Theft: A Reason to Deploy Smart Grid in South Africa' (2019, edn.), pp. 96–101 (2019)
5. Maruf, M.H., Haq, M.A.u., Dey, S.K., Al Mansur, A., and Shihavuddin, A.S.M.: 'Adaptation for sustainable implementation of smart grid in developing countries like Bangladesh. Energy Rep. **6**, 2520–2530 (2020)
6. Mohsin, A., et al.: Energy theft detection in smart grids: taxonomy, comparative analysis, challenges, and future research directions. IEEE/CAA J. Automat. Sin. **9**(4), 578–600 (2022)
7. Kgaphola, P.M., Marebane, S.M., and Hans, R.T.: Electricity theft detection and prevention using technology-based models: a systematic literature review. In: Editor (Ed.)^(Eds.): 'Book Electricity Theft Detection and Prevention Using Technology-Based Models: A Systematic Literature Review' (2024, edn.), pp. 334–350 (2024)
8. Astronomo, J., Dayrit, M.D., Edjic, C., and Regidor, E.R.T.: Development of electricity theft detector with GSM module and alarm system. In: Editor (Ed.)^(Eds.): 'Book Development of Electricity Theft Detector with GSM Module and Alarm System' (2020, edn.), pp. 1–5 (2020)
9. Tanwar, S., et al.: GrAb: a deep learning-based data-driven analytics scheme for energy theft detection. Sensors **22**(11), 1–23 (2022)
10. Haq, E.U., Pei, C., Zhang, R., Jianjun, H., Ahmad, F.: Electricity-theft detection for smart grid security using smart meter data: a deep-CNN based approach. Energy Rep. **9**, 634–643 (2023)
11. Kumarana, K., Ananthib, N., Saranyac, G., Priyadharshinid, S., Thiviyabalae, T., Vaishnavif, K.: Power theft detection and alert system using IOT. Turk. J. Comput. Math. Educ. **12**(9), 12–17 (2021)

12. Dobrić, G., Žarković, M.: Non-technical losses detection in power system. J. Comput. Forensic Sci. **1**(1), 17–28 (2022)
13. Yousif, A., Bashir, M.B., Ali, A.: An evolutionary algorithm for task clustering and scheduling in IoT edge computing. In: Book An Evolutionary Algorithm for Task Clustering and Scheduling in IoT Edge Computing (2024). https://doi.org/10.1109/ACCESS.2020.3011503
14. Li, J., et al.: A secured framework for SDN-based edge computing in IoT-enabled healthcare system. IEEE Access 8, 135479–135490 (2020). https://doi.org/10.1109/ACCESS.2020.3011503
15. Jin, X., Li, L., Dang, F., Chen, X., Liu, Y.: A survey on edge computing for wearable technology. Digit. Sign. Process. **125**, 1–18 (2022)
16. Mhlanga, D., Moloi, T.: 'COVID-19 and the digital transformation of education: what are we learning on 4IR in South Africa?. In Editor (Ed.)^(Eds.): 'Book COVID-19 and the Digital Transformation of Education: What Are We Learning on 4IR in South Africa?' (2020, edn.), pp. 1–11 (2020)
17. Ballal, M.S.: Online electrical energy theft detection and prevention scheme for direct hook-line activity. J. Inst. Eng. (India) Series B, **102**(5), 1007–1018 (2021)
18. Rajasekaran, A., Ajay, S.P., Bhanu, T.P., Swetha, P.P.G.S.: IoT Based power detection. Int. J. Scientific Dev. Res. **5**, 249–253 (2020)
19. Engelbrecht, J., Hanck, G.P., Osifeko, M.O.: Design and implementation of an electrical tamper detection system. In: IECON Proceedings (Industrial Electronics Conference), vol. 1, pp. 2952-2957 (2019)
20. Mohanty, S., Iqbal, M., Thampi, P.: Controlling and monitoring of power theft using Internet of Things. In: 2021 International Conference on Design Innovations for 3Cs Compute Communicate Control (ICDI3C), pp. 111–115 (2021)
21. Dharani, A., Kumar, S., Patil, P.: Object detection at edge using TinyML models. SN Comput. Sci. **5**, 1–6 (2023)
22. Alaa, R., Hussein, E., Al-Libawy, H.: Object detection algorithms implementation on embedded devices: challenges and suggested solutions. Kufa J. Eng. **15**, 148–169 (2024)

Optimization of Transmission Line Switching for Improving Power System Small Signal Stability

AG Oduor[1(✉)], JL Munda[1], AO Akumu[1], and LO Otieno[2]

[1] Department of Electrical Engineering, Tshwane University of Technology, Pretoria 0183, South Africa
alloyce32@gmail.com, {mundajl,akumuao}@tut.ac.za

[2] Department of Electrical and Electronic Engineering, Dedan Kimathi University of Technology, Nyeri 10143, Kenya
luke.otieno@gmail.com

Abstract. Voltage control in power systems can be accomplished by switching off some transmission lines and transferring their loading to the remaining lines. This is normally done during high voltage situations caused by low demand. By so doing, the lines remaining in service will operate above their surge impedance loading thereby consuming reactive power and hence lowering voltage to desired levels. While this operational methodology could be effective, it may introduce or sustain small signal instability, manifested as low frequency oscillations (LFO), which when not adequately damped, can endanger system security and lead to widespread outages in case of disturbances. This paper has evaluated six different multi-objective evolutionary algorithms (MOEAs) that can be used to optimize transmission line switching to concurrently improve small signal stability (SSS) during voltage control. By eigenvalue analysis, the algorithms were tested on the IEEE-30 and IEEE-118 bus models and the results indicate an improvement in SSS. Among the tested algorithms, multi-objective particle swarm optimizer with crowding distance (MOPSO-CD) was found to converge fastest, while competitive mechanism based multi-objective particle swarm optimizer (CMOPSO) yielded better results at moderate convergence rates.

Keywords: transmission line switching · small signal stability · low frequency oscillations · algorithms · eigenvalue · damping ratio

1 Introduction

Power system operations involve switching off transmission equipment to: a) improve voltage profiles, b) enhance transfer capacity, c) benefit from topology change during emergency conditions and d) for cutting costs by incorporating transmission switching into the optimized dispatch [1]. Selective switching off some transmission lines therefore is one of the ways of voltage control in power

T. Zielinska (Ed.): PEAE 2025, LNNS 1679, pp. 153–161, 2026.
https://doi.org/10.1007/978-3-032-08221-3_14

systems. Transmission line switching is also common in power systems operations due to routine maintenance requirements, faulty conditions, economic considerations and power reliability concerns.

While these intentional actions could be effective especially in the control of high voltages, they do change the equilibrium point and hence the modal properties at which a power system was linearized [2] and may create other undesirable results including small signal instability (in form of low frequency oscillations (LFO)) that could impair system security if they are not adequately damped. Maintenance activities that involve switching off transmission lines in regions with weak grid structure increases the risk of poorly-damped LFO [3]. Loading and load distribution in the system also contributes to low frequency oscillatory problems that need to be carefully assessed [4].

Power systems in the developing world are undergoing evolution by deploying converter-based generation resources (mostly renewable energy technologies) and very long transmission line interconnections that by their nature, introduce LFOs under different loading conditions. Most of the control centers in the developing world still rely only on the supervisory control and data acquisition (SCADA) systems which only display root mean square values and do not have the capability for processing and displaying phasors for appropriate remedial action by system operators in the event of undamped oscillations.

While some transmission line switching negatively impact on small signal stability (SSS), there are however some reconfigurations through switching that result in improved damping ratio [5]. Therefore, there is need for an operational methodology which optimizes transmission line switching [6] and also raises the short circuit ratio with the overall objective of improving SSS [7].

This study has therefore performed a rigorous evaluation of the various multi-objective optimization algorithms through modal analysis and proposed one that is robust and yielded the most superior results. During simulations, each algorithm specifies which transmission lines are to be switched off for voltage control, but constrained by SSS criteria.

2 Optimization Approach

There are numerous multi-objective evolutionary algorithms (MOEAs) in literature. Finding a suitable MOEA for a given multi-objective problem (MOP) requires the evaluation of a number of MOEAs for the MOP. For this work, we selected six MOEAs, by considering;

i. types of evolutionary algorithms, such as genetic algorithms (GAs), differential evolution (DE), particle swarm optimization (PSO), and evolutionary strategies (ES)
ii. selection evaluation mechanisms, such as dominance-based approaches and decomposition-based approaches.

In particular, the six selected algorithms compared in this research work are: non-dominated sorting genetic algorithm III (NSGA-III) [8]; competitive mechanism based multi-objective particle swarm optimizer (CMOPSO) [9]; multi-objective particle swarm optimization with crowding distance (MOPSO-CD) [10]; generalized DE (GDE3) [11]; MOEA based on decomposition and covariance matrix evolution strategy (MOEA/D-CMA) [12] and MOEA based on decomposition and DE (MOEA/D-DE) [13]. NSGA-III, CMOPSO, MOPSO-CD and GDE3 are representative of algorithms using dominance-based approaches to select solutions while MOEA/D-DE and MOEA/D-CMA are representative of decomposition-based approaches. Selection evaluation also incorporates mechanisms for collecting diverse solutions from the Pareto front (PF). MOEA/D-CMA is the algorithm that represents ES, while PSO, GA, and DE are represented by the algorithms with corresponding names. The bases of comparison of the selected algorithms are: the final solutions and the number of iterations required for convergence.

The implementations of all the selected algorithms can; a) handle integer decision variables and, b) optimize MOPs. An open-source evolutionary optimization platform called PlatEMO was used to solve the MOP [14]. The simulations are carried out using MATLAB and power systems analysis toolbox (PSAT) [15].

While implementing transmission line switching, the final objectives were to: a) minimize the maximum real part of the eigenvalue and b) maximize the minimum damping ratio of linearized IEEE-30 and IEEE-118 bus models through a multi-objective optimization process. The MOP was formulated as follows:

a. **Objectives**

1. Minimizing the maximum real part of eigenvalues, σ_{max} (to improve SSS)

$$\underset{\mathbf{l}}{\text{minimize}} \quad \sigma_{\max} \tag{1}$$

   where $\mathbf{l}$ is the vector listing the lines that are switched off in a network.

2. Maximizing the minimum damping ratio, ζ_{min} (to improve SSS)

$$\underset{\mathbf{l}}{\text{minimize}} \quad -\zeta_{\max} \tag{2}$$

b. **Constraints**

1. The minimum damping ratio, $\zeta_{\min}$, must be non-negative (to ensure SSS is not violated)

$$\zeta_{\min} \geq 0 \tag{3}$$

2. The maximum real part of the eigenvalues, $\boldsymbol{\sigma_{\max}}$, must be in the left half of the s-plane (to ensure SSS is not violated)

$$\sigma_{\max} \geq 0 \tag{4}$$

3. Range of allowable bus voltage regulation, V_r (to ensure voltage stability is not violated)

$$-0.1 \leq V_r \leq 0.1 \tag{5}$$

4. Range of number of lines to be switched off, l (to avoid violation of $N-2$ contingency criterion and transfer capacity)

$$1 \leq l \leq 2 \tag{6}$$

5. Limit of voltage phase angle difference, δ_i, between either ends of any line i in the system (to avoid angle instability)

$$\mid \delta_i \mid \leq 30^\circ \tag{7}$$

6. The number of weakly connected components (WCC) in the system, n_{WCC}, after reconfiguration must be equal to 1 (to avoid islanding)

$$n_{\mathrm{WCC}} = 1 \tag{8}$$

3 Simulations

The study was done in four stages, the first three being carried out on the IEEE-30 bus model, while the fourth was undertaken on IEEE-118 bus system. The tests were conducted on diverse systems to justify generalization of the algorithmic results and support the conclusions.

The first stage was done only to reveal which transmission lines each optimization algorithm would specify for switching off for voltage control. These lines would be used to determine at which buses the loads would be adjusted during the later stages of the study. The buses are those that are electrically closer to these affected lines.

The second stage was the base case where there was no optimization and no switching. No algorithm was used and a basic eigenvalue analysis was done using PSAT. Loads were adjusted (at 0.0 pu, 0.25 pu, 0.5 pu, 0.8 pu, 1.0 pu and 1.2 pu) at the buses that were already identified during the first stage of simulations. The results would later be used for comparison of small signal stability index (damping ratios and eigenvalues) between situations where there is no switching and instances of optimized switching.

The third stage of the study was done by simulating (by all the algorithms) the cases when the specified lines were switched off, while the loads at the identified buses are adjusted as above. At this stage, since it is expected that transmission line switching affects power flow distribution and system reliability, the impact of switching off some transmission lines by the algorithm was also studied on IEEE-30 bus model from both voltage stability and angle stability perspectives to represent the real situations in power system operations.

The fourth stage of the study was done by simulating (by all the algorithms) the cases when the specified lines were switched off in IEEE-118 bus system, while the loads at the identified buses are adjusted.

4 Results

Results of the damping ratios and eigenvalues at 1.0pu loading (which is the same for all the buses) for the IEEE-30 bus system are presented in Table 1. Table 1 also indicates the results when the first objective function is prioritized in the optimization and also when the second objective function is prioritized. The table further compares the number of iterations/generations each algorithm takes to converge (Gens). Table 2 presents a detailed optimized results for all the algorithms at the various bus loadings including PSAT results (where there is no switching). Figures 1(a) to (f) present the evolution of the solutions at selected generations for each of the algorithms at 1.0 pu bus loading.

Table 3 presents the IEEE-30 bus case results of voltage sensitivity at some five buses (that are impacted most by the optimized switching) together with the corresponding phase angles as the loading at bus 4 was being varied after optimized switching. Bus 4 was selected because it is directly linked to line 3 which has been switched off by the MOPSO-CD optimizer.

Table 4 presents a comparison of the eigenvalues ($\sigma_{\max}$) and damping ratios ($\zeta_{\min}(\%)$) results at 1.0 pu loading between the optimization algorithms for the IEEE-118 bus model. The table also shows the lines that each optimizer specifies for switching off.

Table 1. Results from the selected optimizers IEEE-30 bus model.

Algorithm	Gens	Prioritization					
		Objective 1			Objective 2		
		$\sigma_{\max}$	$\zeta_{\min}$ (%)	lines	$\sigma_{\max}$	$\zeta_{\min}$ (%)	lines
CMOPSO	6	−0.07126	2.13205	8	−0.07034	2.13718	3
MOPSO-CD	1	−0.07126	2.13205	8	−0.07034	2.13718	3
NSGA-III	16	−0.07128	2.13258	8, 16	−0.07034	2.13718	3
GDE3	37	−0.07126	2.13205	8	−0.07034	2.13718	3
MOEA/D-CMA	36	−0.07126	2.13205	8	−0.07034	2.13718	3
MOEA/D-DE	27	−0.07126	2.13205	8	−0.07034	2.13718	3

5 Discussion

a. All the algorithms yield almost similar results that improve small signal stability as can be seen in both Tables 1 and 4 together with the pareto fronts in Figs. 1(a) to (f).
b. In Table 4, it can be seen that while the base case has positive eigenvalue, some of the algorithms improve SSS by yielding non-negative eigenvalues and damping ratios.
c. In both the IEEE-30 and IEEE-118 bus systems, MOPSO-CD converges fastest as indicated in Tables 1 and 4 together with Fig. 1(b). However, it might be trapped in a local optimum as results in Table 4 show, since other algorithms yield better solutions of eigenvalues and damping ratio.

Table 2. Results at various bus loadings IEEE-30 bus model. Loading variations at 0.0 pu, 0.25 pu, and 0.8 pu showed similar patterns but have been omitted due to space limitation.

Loads	Reconfiguration	Load (pu)					
		0.50		1.00		1.20	
		ζ (%)	σ	ζ (%)	σ	ζ (%)	σ
7	No switching	2.13298	−0.07134	2.12660	−0.07110	2.12390	−0.07101
	Others	2.14486	−0.07132	2.13718	-0.07126	2.13374	−0.07123
	NSGA-III	2.14486	−0.07134	2.13718	-0.07128	2.13374	−0.07125
8	No switching	2.14421	−0.07176	2.12660	-0.07110	2.11970	−0.07085
	Others	2.15412	−0.07196	2.13718	−0.07126	2.13003	−0.07096
	NSGA-III	2.15412	−0.07198	2.13718	−0.07128	2.13003	−0.07098
21	No switching	2.12515	−0.07104	2.12660	−0.07110	2.12706	−0.07113
	Others	2.13681	−0.07120	2.13718	−0.07126	2.13720	−0.07127
	NSGA-III	2.13681	−0.07123	2.13718	−0.07128	2.13720	−0.07129
29	No switching	2.12926	−0.07120	2.12660	−0.07110	2.12521	−0.07107
	Others	2.13765	−0.07133	2.13718	−0.07126	2.13713	−0.07123
	NSGA-III	2.13765	−0.07136	2.13718	−0.07128	2.13713	−0.07125
30	No switching	2.15981	−0.07211	2.12660	−0.07110	2.11488	−0.07076
	Others	2.16137	−0.07213	2.13718	−0.07126	2.12982	−0.07096
	NSGA-III	2.16212	−0.07216	2.13718	−0.07128	2.12982	−0.07098

d. The results in Tables 1 and 4 show that the best eigenvalues and damping ratios are yielded by CMOPSO, MOEA/D-CMA, and MOEA/D-DE. However, among the three algorithms, CMOPSO's convergence is the fastest.
e. NSGA-III and GDE3 fail to yield feasible solutions starting from an unstable IEEE-118 base case. This could be because of the algorithms' exploratory behavior in a space where the likelihood of getting infeasible solutions is higher than the likelihood of getting feasible solutions. Performance of these algorithms for this application can perhaps be improved by careful tuning of their parameters.
f. Of all the six algorithms, MOPSO-CD converged after the least number of generations as can be seen in Tables 1 and 4.
g. Optimized switching generally improves SSS. As can be seen in Table 2, they yield superior results compared to the situations when there is no switching off of lines (i.e. when all lines are in service - base case). This is consistent even with load variations in the system. When any bus is loaded at 1.0 pu, the damping ratio and real part of the eigenvalue from MOPSO-CD are greater than that of PSAT base case by 0.5% and 0.23% respectively.
h. SSS is quite sensitive to system loading as shown in Table 2. With very high loading, the system tends towards small signal instability (damping ratios and real part of eigenvalues become poorer), but it improves with low load-

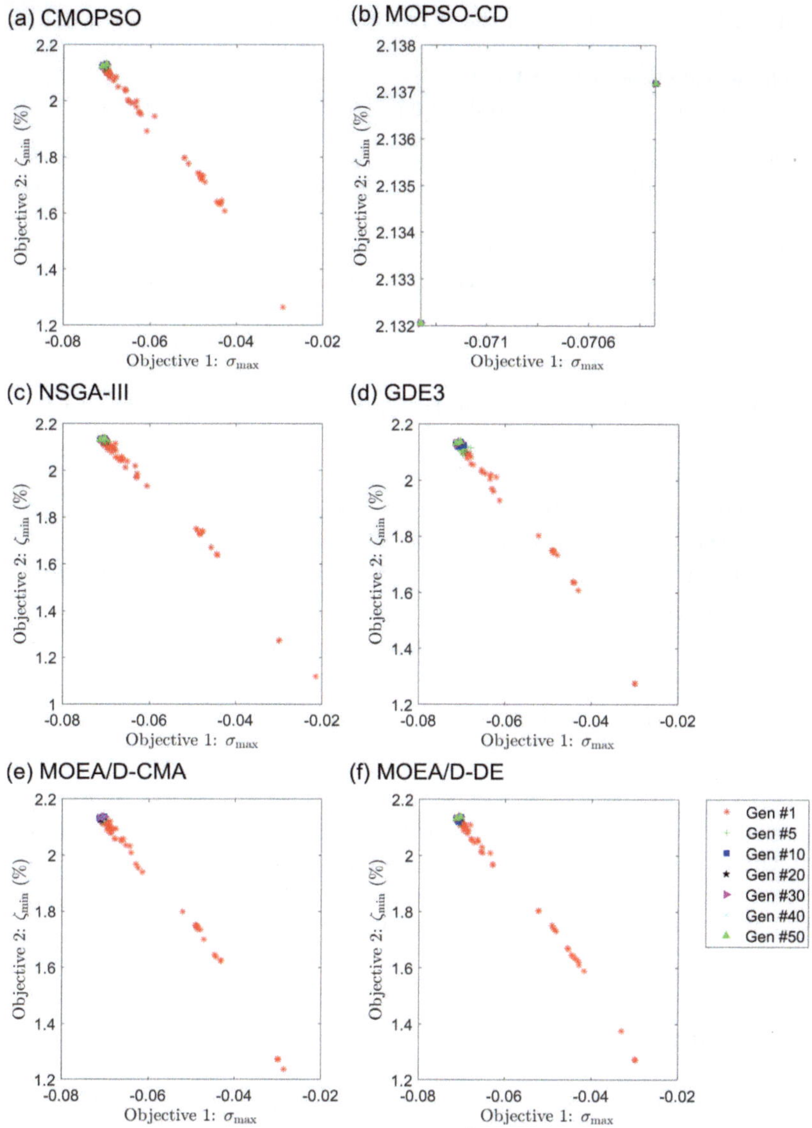

Fig. 1. Convergence of solutions with the number of generations.

ing levels except for a few cases under bus 21. This is consistent with the expectation.

i. Irrespective of which bus the load is being adjusted (Table 2), each algorithm yields the same results at 1.0 pu of bus loading which is also consistent with the known or expected fact.
j. As expected, the remaining lines in service are loaded more compared to base case, which has effect in voltage drop and increase in phase angle difference.

Table 3. Voltages at bus 4 in the IEEE-30 model after optimized switching.

Bus No.	Voltages and phase angles at Bus 4					
	Load = 0.5 pu		Load = 1.0 pu		Load = 1.2 pu	
	voltage (pu)	angle (degrees)	voltage (pu)	angle (degrees)	voltage (pu)	angle (degrees)
8	0.9498	−5.52	0.949	−5.82	0.9486	−5.94
19	0.9546	−8.46	0.9543	−8.78	0.9543	−8.9
20	0.9573	−8.46	0.9571	−8.78	0.957	−8.91
29	0.9507	−4.56	0.9507	−4.87	0.9507	−4.99
30	0.9081	−4.45	0.9081	−4.76	0.9081	−4.88

Table 4. Results from the selected optimizers IEEE-118 bus model.

Case	$\sigma_{\max}$	$\zeta_{\min}$ (%)	Lines off	Gens
Base case	0.4206	−100	–	–
NSGAIII	No feasible solution	No feasible solution	–	–
CMOPSO	−0.0106	1.7934	132, 22	26
MOPSO-CD	−0.0106	1.74095	143. 172	4
GDE3	No feasible solution	No feasible solution	–	–
MOEA/D-CMA	−0.0106	1.7934	132, 22	39
MOEA/D-DE	−0.0106	1.7934	132. 22	50

But as it can be seen in Table 3, these parameters have not been violated by optimized switching since no bus voltage drops below 0.9 pu, while phase angles are much below 30°. The lowest voltage is at bus 30 which remains at 0.9081 pu, while the largest phase angle is −8.91° at bus 20 when the bus is loaded at 1.2 pu.

6 Conclusion

This study has employed eigenvalue analysis to evaluate the multi-objective algorithms that are best suited for transmission line switching especially for voltage control while improving system SSS. The algorithms have been tested on both the IEEE-30 and IEEE-118 bus models and results indicate that when a methodology involving the use of these algorithms is implemented, SSS is improved and it can be used by power system control engineers to improve small signal oscillatory performance while undertaking grid control and operations. When iteration time is of priority, then MOPSO-CD is recommended for implementation while if a balance is to be struck between iteration time and improvement of SSS, then CMOPSO is recommended.

Finally, in order to generalize the outcomes of this study, further tests on other test systems can be carried out.

References

1. Hedman, K.W., O'Neill, R.P., Fisher, E.B., Oren, S.S.: Optimal transmission switching with contingency analysis. IEEE Trans. Power Syst. **24**(3), 1577–1586 (2009)
2. Setareh, M., Parniani, M., Aminifar, F.: An analytic methodology to determine generators redispatch for proactive damping of critical electromechanical oscillations. Int. J. Electr. Power Energy Syst. **123**, 106301 (2020)
3. Wang, C., et al.: Low frequency oscillation characteristics of East China power grid after commissioning of Huai-Hu ultra-high voltage alternating current project. J. Modern Power Syst. Clean Energy **3**(3), 332–340 (2015)
4. Dheer, D.K., Kulkarni, O.V., Doolla, S., Rathore, A.K.: Effect of reconfiguration and mesh on small signal stability margin of a droop-based islanded microgrid. In: 2016 IEEE International Conference on Power Electronics, Drives and Energy Systems (PEDES), pp. 1–6. IEEE (2016)
5. Chen, Y., Fuller, J., Diao, R., Zhou, N., Huang, Z., Tuffner, F.: The influence of topology changes on inter-area oscillation modes and mode shapes. In: 2011 IEEE Power and Energy Society General Meeting, pp. 1–7. IEEE (2011)
6. Hedman, K.W., Oren, S.S., O'Neill, R.P.: Optimal transmission switching: economic efficiency and market implications. J. Regul. Econ. **40**, 111–140 (2011)
7. Dong, S., Gao, N., Tan, J., O'Neill, B.: Power system oscillations in high renewable power systems. Accessed 04 Feb 2025. https://youtu.be/fJ6JlRtQFhc?si=80jRTVWySDi84MIw
8. Deb, K., Jain, H.: An evolutionary many-objective optimization algorithm using reference-point-based nondominated sorting approach, part i: solving problems with box constraints. IEEE Trans. Evol. Comput. **18**(4), 577–601 (2013)
9. Zhang, X., Zheng, X., Cheng, R., Qiu, J., Jin, Y.: A competitive mechanism based multi-objective particle swarm optimizer with fast convergence. Inf. Sci. **427**, 63–76 (2018)
10. Raquel, C.R., Naval, P.C.: An effective use of crowding distance in multiobjective particle swarm optimization. GECCO '05, pp. 257–264. Association for Computing Machinery, New York, NY, USA (2005)
11. Kukkonen, S., Lampinen, J.: GDE3: the third evolution step of generalized differential evolution. In: 2005 IEEE Congress on Evolutionary Computation, vol. 1, pp. 443–4501 (2005)
12. Zapotecas-Martínez, S., Derbel, B., Liefooghe, A., Brockhoff, D., Aguirre, H.E., Tanaka, K.: Injecting CMA-ES into MOEA/D. In: Proceedings of the 2015 Annual Conference on Genetic and Evolutionary Computation. GECCO '15, pp. 783–790. Association for Computing Machinery, New York, NY, USA (2015)
13. Li, H., Zhang, Q.: Multiobjective optimization problems with complicated pareto sets, MOEA/D and NSGA-II. IEEE Trans. Evol. Comput. **13**(2), 284–302 (2009)
14. Tian, Y., Cheng, R., Zhang, X., Jin, Y.: PlatEMO: a MATLAB platform for evolutionary multi-objective optimization [Educational Forum]. IEEE Comput. Intell. Mag. **12**(4), 73–87 (2017)
15. Milano, F.: An open source power system analysis toolbox. IEEE Trans. Power Syst. **20**(3), 1199–1206 (2005)

Power Systems Distributed Inertia Estimation

Louwrence Ngoma(✉), Josiah Munda, and Yskandar Hamam

Tshwane University of Technology, Pretoria 0183, South Africa
{ngomaJL,mundaJL,hamamY}@tut.ac.za

Abstract. Traditional inertia estimation techniques, which assume a fixed and centralized inertia distribution, are becoming less effective due to the spatial variability introduced by renewable energy sources. This study proposes an inertia estimation method using a polynomial fitting approach to improve estimation accuracy by analyzing frequency and power deviations at multiple system buses. The method is tested on the Kundur two-area system and the IEEE 39-bus system, where disturbances are introduced at different locations to assess spatial inertia variations. The results indicate that estimation errors are highest at the faulted buses and decrease progressively with distance from the disturbance.

Keywords: Distributed Inertia · Inertia Estimation · Polynomial Fit · Spatial Variability

1 Introduction

The stability of power systems is crucial for ensuring a reliable electricity supply, particularly in the presence of disturbances that cause frequency deviations. Inertia, inherently provided by the rotating masses of synchronous generators, plays a fundamental role in mitigating these deviations by resisting sudden changes in system frequency [1]. Traditional power systems assume a centralized and relatively uniform distribution of inertia. However, the increasing penetration of renewable energy sources has introduced spatial variations that significantly impact frequency stability and system response to disturbances [2]. Accurately estimating inertia is essential for power system operators and planners, as it enables the development of effective control strategies to maintain grid stability.

Conventional inertia estimation techniques, such as those based on system identification or swing equation approaches, often assume a uniform inertia distribution across the network [3]. However, as demonstrated by [4], these methods become less accurate when dealing with spatial variations. Measurement-based approaches using phasor measurement units (PMUs) have improved inertia estimation accuracy, yet challenges remain in capturing localized variations in inertia [5, 6]. In [7], the authors proposed a method using synchrophasor data for inertia estimation in the Great Britain power system. [8] estimated inertia based on an iterative equation error system identification approach, but their approach was found to be sensitive to network topology changes. Area inertia estimation approaches, as discussed by [9, 10], rely on the aggregate power

T. Zielinska (Ed.): PEAE 2025, LNNS 1679, pp. 162–171, 2026.
https://doi.org/10.1007/978-3-032-08221-3_15

disturbances and area center of inertia (COI) frequency to derive localized inertia characteristics. These methods highlight the complexities associated with estimating inertia, especially in systems with mixed generation types and configurations. Despite the variety of inertia estimation methods proposed, many existing approaches often neglect the impact of disturbance location on inertia estimation accuracy. The spatial variation in inertia due to localized disturbances remains a critical yet underexplored challenge in power system analysis.

To address this gap, this study proposes an inertia estimation method that uses polynomial fitting to analyze frequency and power deviations at multiple system buses. Unlike conventional methods, the proposed approach explicitly accounts for the spatial dependence of inertia, providing a more accurate and practical estimation technique for modern power systems.

The remainder of the paper presents the methodology of the polynomial fitting inertia estimation method in Sect. 2. Section 3 discusses case studies on the Kundur two-area system and IEEE 39-bus system, evaluating the method's effectiveness in capturing spatial variations in inertia. Section 4 assesses the accuracy of the proposed approach using the system average inertia (SAI) method. Finally, Sect. 5 summarizes the key results and outlines future research directions.

2 Power system Inertia Estimation

In power systems, large disturbances cause power imbalances that trigger inertial responses before governor actions take effect. The system's frequency response during this inertial response is described by:

$$\frac{2SH}{f_0}\frac{df}{dt} = P_m - P_e \tag{1}$$

where f_0 is the nominal system frequency, P_m is the mechanical power of the synchronous generator, P_e is the electrical power of the synchronous generator, S is the nominal apparent rated power of the generator, f is the measured frequency, and H represents the generator's inertia constant.

Assume $P_m = 0$, since generator governors have a longer response time than the inertial stage, the deviation in electrical power primarily drives frequency dynamics.

$$H = -\frac{\Delta P_e}{2S} \times \frac{f_0}{\frac{df}{dt}} \tag{2}$$

To improve the accuracy of inertia estimation, the variable order polynomial fitting technique is applied to the frequency response measured at the point of disturbance. Figure 1 shows the steps of the proposed inertia estimation method.

Step 1: Perform a load flow analysis to determine the pre-disturbance operating conditions.
Step 2: Collect frequency and power data from multiple buses using PMUs.
Step 3: Apply polynomial fitting to extract frequency Δf and power ΔP deviations.

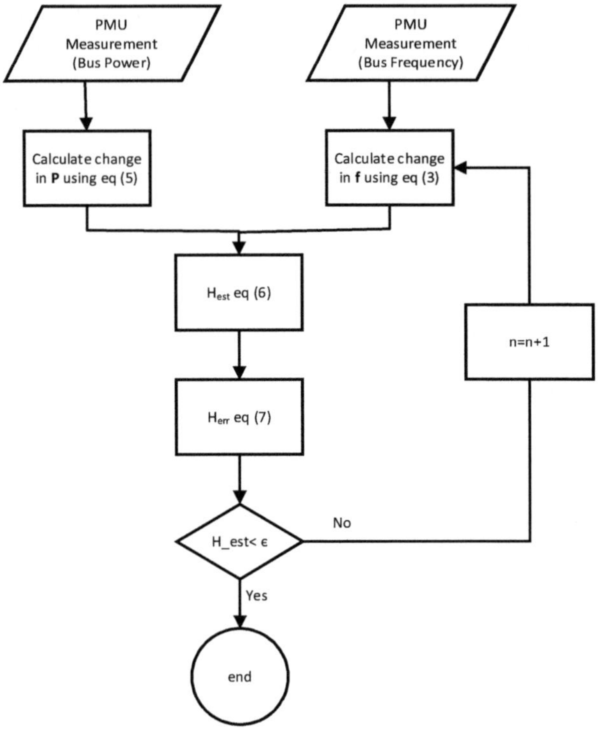

Fig. 1. Inertia estimation

The polynomial representation of frequency deviation Δf over time is given as

$$\frac{\Delta f(t)}{f_0} = A_n t^n + A_{n-1} t^{n-1} + A_{n-2} t^{n-2} + \cdots + A_1 t + A_0 \tag{3}$$

The polynomial fit of power deviation ΔP is expressed as

$$\frac{\Delta P(t)}{P_0} = B_k t^k + B_{k-1} t^{k-1} + B_{k-2} t^{k-2} + \cdots + B_1 t + B_0 \tag{4}$$

The total electrical power $P_e(t)$ is

$$P_e(t) = P_0 + \Delta P(t) = P_0\left(1 + B_k t^k + B_{k-1} t^{k-1} + \cdots + B_1 t + B_0\right) \tag{5}$$

where Δf is the deviation of measured frequency from the nominal frequency f_0. ΔP is the deviation of measured power from the steady state value P_0 before a fault occurs.

Step 4: Use the polynomial coefficients, from the previous step, in Eq. (6) to compute the inertia constant H_{est}.

A_1 and B_1 are the key coefficients obtained through polynomial curve fitting in step 3. The estimated inertia constant is then derived as:

$$H_{est} = \frac{B_1}{2SA_1} \tag{6}$$

Step 5: Using Eq. (6), compare estimated inertia with actual machine values. If the error exceeds a set threshold, increment the polynomial order to refine the estimation iteratively.

The percentage error in inertia estimation is calculated as:

$$H_{err} = \left| \frac{H_{est} - H_{actual}}{H_{actual}} \right| \quad (7)$$

3 Case Studies

The proposed power system inertia estimation method is evaluated on two test systems, the Kundur two-area test system and the IEEE 39-Bus test system. The systems are subjected to power imbalance events at different generator buses to assess how the proposed method performs in detecting spatial variations in inertia and estimating overall system stability. MATLAB software is used for system simulation and inertia estimation.

3.1 Kundur Two-Area Test Power System

Figure 2 shows the Kundur two-area test power system, which consists of four generators, ten buses, and two loads. A three-phase fault lasting 200 ms is introduced at generator buses to analyze frequency response and inertia estimation accuracy.

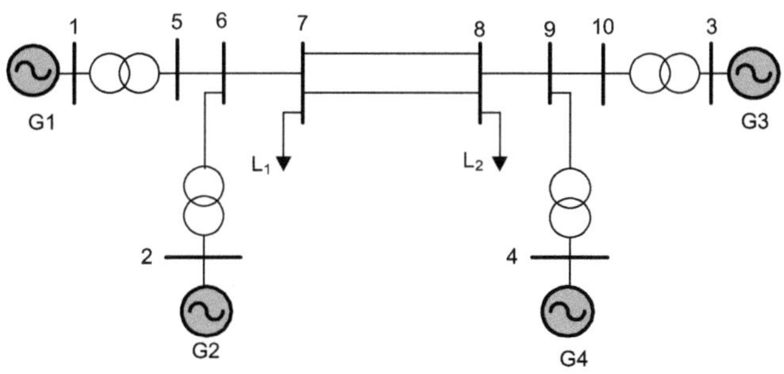

Fig. 2. Kundur two-area test power system

The frequency deviations following a fault at Bus 1 are shown in Fig. 3. Buses closer to the disturbance exhibit higher deviations, confirming the spatial variation in inertia distribution. The largest deviation is observed at Bus 1, demonstrating that inertia estimation is more challenging at faulted locations due to significant frequency disturbances. Table 1 presents the estimated inertia values, showing the highest error of 3.877% at Bus 1, with a gradual decrease at Buses 2, 3, and 4. This pattern highlights that generators farther from the disturbance experience lower estimation errors, indicating a more stable frequency response.

Similarly, when a fault occurs at Bus 4, the frequency deviations follow the same trend as shown in Fig. 4. The highest error, 4.032%, is recorded at Bus 4, as shown in

Table 2. The error reduces progressively at buses farther from the fault, reinforcing the observation that spatial variations in inertia estimation accuracy depend on proximity to the disturbance. These results confirm that the method successfully captures localized inertia variations, making it a valuable tool for grid operators in stability assessments.

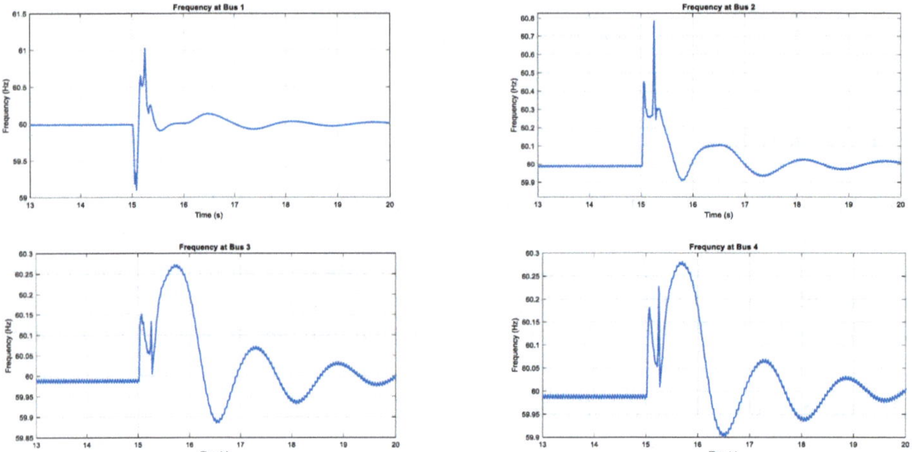

Fig. 3. Fault at Bus 1

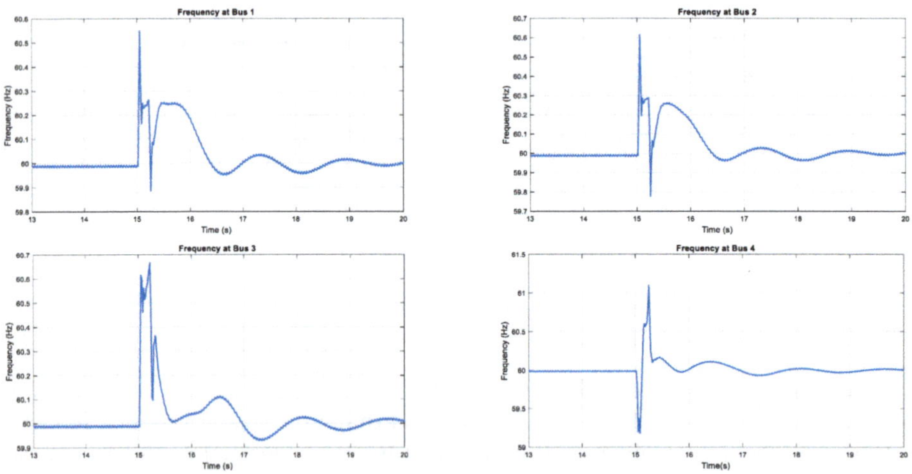

Fig. 4. Fault at Bus 4

Although the Kundur system provides valuable insights into a smaller network, real-world power grids are often larger and more interconnected. To further validate the method's effectiveness, the analysis is extended to the IEEE 39-bus system.

Table 1. Inertia estimation when there is a fault at bus 1

Bus	H_{gen}	H_{est}	% Error
1	6.5	6.752	3.877%
2	6.5	6.569	1.062%
3	6.175	6.163	0.194%
4	6.175	6.171	0.065%

Table 2. Inertia estimation when there is a fault at bus 4

Bus	H_{gen}	H_{est}	% Error
1	6.5	6.537	0.569%
2	6.5	6.443	0.877%
3	6.175	6.297	1.976%
4	6.175	6.424	4.032%

3.2 IEEE 39 Bus Test Power System

Figure 5 shows a more complex network with ten generators representing a realistic model of a large interconnected power system. A three-phase fault lasting 200 ms is introduced at Buses 30 and 31 to assess inertia estimation performance in a large-scale setting. Other buses were also considered, but are not included here due to space limitations.

Tables 3 and 4 show the results for faults at Buses 30 and 31, respectively. The highest estimation errors occur at the faulted buses, with 4.960% at Bus 30 and 4.487% at Bus 31, indicating that disturbances introduce localized measurement uncertainties. Errors decrease at buses farther from the fault, aligning with the trends observed in the Kundur system. The consistency of these trends across different disturbances and system scales confirms the robustness of the proposed method in various operating conditions.

Further analysis reveals that in larger systems like the IEEE 39-bus test system, the impact of localized disturbances is more distributed due to the presence of multiple generators and interconnected networks. The proposed method effectively accounts for these variations, ensuring accurate inertia estimation. This is particularly important for modern power grids integrating renewable energy sources, where spatial inertia variations play a crucial role in frequency stability.

These results demonstrate that the method effectively captures spatial variations in inertia, providing accurate estimates even in large-scale systems. The spatial dependence of estimation accuracy highlights the need to consider disturbance locations in inertia estimation methods.

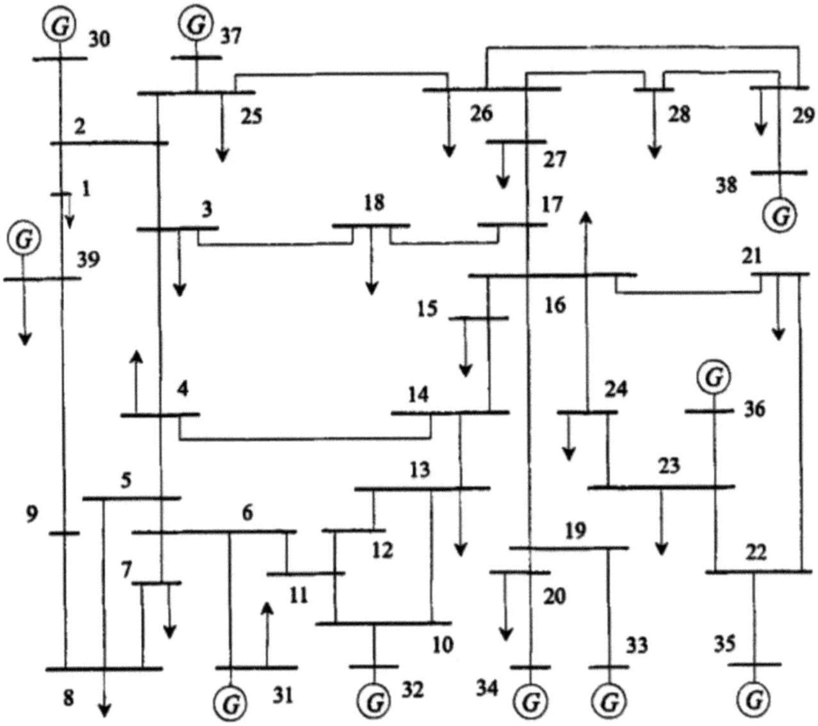

Fig. 5. IEEE 39 Bus test system

Table 3. Inertia estimation when there is a fault at bus 30

Bus	H_{gen}	H_{est}	% Error
30	5.000	5.248	4.960%
31	4.145	4.217	1.737%
32	4.524	4.436	1.945%
33	3.820	3.727	2.435%
34	4.450	4.328	2.742%
35	4.282	4.361	1.845%
36	3.544	3.595	1.439%
37	3.238	3.393	4.787%
38	3.392	3.514	3.597%
39	4.361	4.428	1.536%

Table 4. Inertia estimation when there is a fault at bus 31

Bus	H_{gen}	H_{est}	% Error
30	5.000	5.113	2.260%
31	4.145	4.331	4.487%
32	4.524	4.706	4.023%
33	3.820	3.711	2.853%
34	4.450	4.334	2.607%
35	4.282	4.315	0.771%
36	3.544	3.547	0.085%
37	3.238	3.186	1.606%
38	3.392	3.433	1.209%
39	4.361	4.515	3.531%

4 Performance Evaluation

The performance of the inertia estimation method is assessed using the system average inertia (SAI) calculated from the estimated inertia values in Tables 1, 2, 3, and 4. The SAI values are compared with the actual generator inertia H_{gen} SAI to evaluate accuracy.

4.1 Kundur Two-Area Test System

For the fault at Bus 1, the SAI was calculated using the estimated inertia values from Table 1. Since each generator is rated at 900 MVA, the SAI is computed as follows:

$$H_{SAI} = \frac{\sum_i^N H_{SG,i} S_{B,i}}{\sum_i^N S_{B,i}} \tag{8}$$

As shown in Table 5, the SAI for Bus 1 fault was found to be 6.348 s, while for Bus 4 fault, it was 6.425 s. These values indicate that disturbances at different buses impact the perceived system inertia differently. The actual SAI, calculated using the generator inertia values, was 6.32 s. The deviations between estimated and actual SAI values were 0.44% for Bus 1 fault and 1.66% for Bus 4 fault, as presented in Table 5. The slightly higher error for Bus 4 confirms that spatial variations in inertia influence estimation accuracy.

Table 5. SAI for faults in Kundur system

Fault Location	SAI (s)	Actual SAI (s)	% Error
Bus 1	6.348	6.32	0.44%
Bus 4	6.425	6.32	1.66%

4.2 IEEE 39-Bus Test System

For the IEEE 39-bus system, inertia estimation was evaluated for disturbances at Bus 30 and Bus 31. The computed SAI values are summarized in Table 6. The SAI for Bus 30 fault was 6.414 s, with an error of 1.49%, while for Bus 31 fault, it was 6.385 s, with an error of 1.02%. These results confirm that faulted locations exhibit higher deviations, reinforcing the role of spatial inertia variations.

Table 6. SAI for faults in IEEE 39-Bus system

Fault Location	SAI (s)	Actual SAI (s)	% Error
Bus 30	4.213	4.150	1.52%
Bus 31	4.218	4.150	1.64%

For both faults, the estimated SAI values are very close to the actual system inertia, with a maximum error of 1.64% at Bus 31.

4.3 Comparative Analysis

The proposed polynomial fitting approach offers a more accurate and practical solution for inertia estimation. Traditional methods like the sliding window ROCOF-based technique estimate inertia during large disturbances by calculating the rate of change of frequency over a fixed window (typically 500 ms to 1 s). While simple and fast, this method is highly sensitive to noise and the choice of window size, often producing inaccurate results, particularly near faulted buses where frequency changes are most severe. It also lacks spatial resolution, treating the system as a whole rather than assessing local variations [4, 5]. Ambient data-based methods, such as the ARMAX model, use continuous frequency and power measurements to estimate inertia over longer periods [5]. Though useful for system-wide monitoring, these models do not capture spatial differences and are prone to errors from noise or missing data. In contrast, the proposed method fits variable-order polynomials to frequency deviations right after a disturbance. It smooths out noise, adaptively adjusts the fit for higher accuracy, and estimates inertia at each bus individually, effectively capturing the spatial variations in inertia that traditional uniform distribution assumptions fail to address, particularly showing reduced estimation errors with increasing distance from the faulted buses.

5 Conclusion

This study demonstrates that disturbance location significantly affects inertia estimation accuracy in power systems, with errors highest at faulted buses and decreasing with distance. The proposed polynomial fitting method effectively captures spatial variations in inertia, offering a more accurate alternative to traditional uniform-distribution assumptions. Simulations on the Kundur and IEEE 39-bus test systems show that the model fitting approach provides more accurate inertia estimates compared to sliding window methods and ambient data-based methods. Expanding on these findings, future work should evaluate the computational complexity and scalability of this technique when applied to large modern power systems with diverse renewable energy integration.

Acknowledgment. The authors express their sincere gratitude to the NRF BAAP (National Research Foundation, Black Academic Advancement Programme) for its financial support, which made this work possible.

References

1. Kisinga, D.A.: A virtual synchronous generator control in the power system with high penetration of renewable energies--weak grid!, University of Sheffield (2025)
2. He, C., Geng, H., Rajashekara, K., Chandra, A.: Analysis and control of frequency stability in low-inertia power systems: a review. IEEE/CAA J. Autom. Sin. **11**(12), 2363–2383 (2024)
3. Milano, F., Dörfler, F., Hug, G., Hill, D.J., Verbič, G.: Foundations and challenges of low-inertia systems. In: 2018 power systems computation conference (PSCC), pp. 1–25. IEEE (2018)
4. Prabhakar, K., Jain, S.K., Padhy, P.K.: Inertia estimation in modern power system: a comprehensive review. Electr. Power Syst. Res. **211**, 108222 (2022)
5. Tan, B., Zhao, J., Netto, M., Krishnan, V., Terzija, V., Zhang, Y.: Power system inertia estimation: Review of methods and the impacts of converter-interfaced generations. Int. J. Electr. Power Energy Syst. **134**, 107362 (2022)
6. Dowrah, P., Lian, K.M., Sangma, J.D., Dkhar, F., Affijulla, S.: Estimation techniques for power system inertia: a simulation oriented review. In: 2021 IEEE Mysore Sub Section International Conference (MysuruCon), pp. 36–41. IEEE (2021)
7. Ashton, P.M., Saunders, C.S., Taylor, G.A., Carter, A.M., Bradley, M.E.: Inertia estimation of the GB power system using synchrophasor measurements. IEEE Trans. Power Syst. **30**(2), 701–709 (2014)
8. Gotti, D., et al.: Inertia estimation of a power system area based on iterative equation error system identification. IEEE Trans. Power Syst. **39**(5), 6469–6481 (2024)
9. Tan, B., Peng, J., Gao, N., Zhao, J., Tan, J.: Comparative study of data-driven area inertia estimation approaches on WECC power systems. In: 2024 IEEE Power & Energy Society General Meeting (PESGM), pp. 1–5. IEEE (2024)
10. Swetala, E.G., Kaberere, K.K., Munda, J.K.: Offline estimation of power system inertia distribution. e-Prime-Adv. Electr. Eng. Electr. Energy 100923 (2025)

Research on APU Anti-corrosion Design and Storage

Ma Yang(✉), Li Yanshuang, Gu Cheng, and Xue Weiwei

Shanghai Aircraft Design and Research Institute, Shanghai, China
{mayang,liyanshuang,gucheng,xueweiwei}@comac.cc

Abstract. This paper explains the principle of APU (Auxiliary Power Unit) corrosion from the APU engine and corrosion mechanism. At the same time, the general anti-corrosion design of APU is introduced, explaining that currently anti-corrosion is usually ensured through early anti-corrosion design and subsequent storage programs. This paper provides the impact of corrosion on APU and subsequent recommended measures when APU storage programs are not executed as required. Finally, a specific case study of a flight route is presented.

Keywords: APU · Corrosion · Storage

1 Introduction

APU is an important part of an aircraft, providing air and power to the aircraft in emergency situations on the ground or in the air. With the development of APU, especially in the civil aviation field, the requirements for safety, economy, and maintainability are becoming increasingly high.

The APU body is usually composed of an air collecting chamber, a compressor, a combustion chamber and a turbine. The air enters the compressor through the air collecting chamber, and is compressed by the compressor into the combustion chamber, forming a high-temperature and high-pressure gas driven turbine to rotate, and converting heat energy into mechanical energy. At the same time, part of the compressed gas of the compressor is provided to the aircraft air conditioning system and engine starting through the air entrain, and the turbine rotor shaft drives the APU generator through the reducer to provide electrical energy for the aircraft.

With the development of science and technology, APU is required to be simple in structure, light in weight, easy to maintain and low in cost [1]. Considering maintenance and cost requirements, the materials currently used in APUs may corrode if exposed to moisture for a long time, especially compressor cartridges and turbine guides. The corrosion of compressor blades in naval aviation aircraft engines is caused by the warm, humid, and oceanic atmosphere [2]. Especially in coastal areas (with high humidity, high salt content, and industrial waste gas), engines will be affected by multiple factors, leading to corrosion [3]. The corrosion propagation speed is fast, and the fault is relatively hidden [4]. Therefore, it is necessary to combine the maintenance and use of APU to formulate reasonable storage procedures for APU anticorrosion.

T. Zielinska (Ed.): PEAE 2025, LNNS 1679, pp. 172–178, 2026.
https://doi.org/10.1007/978-3-032-08221-3_16

Starting from the underlying mechanism, this paper introduces the corrosion mechanism of APU and the anti-corrosion means.

2 Introduction of Corrosion Mechanism

Corrosion is a spontaneous change in the state and properties of a substance, which is a chemical and electrochemical reaction determined by the second law of thermodynamics and is an inevitable natural phenomenon.

Corrosion in a broad sense refers to the destruction or deterioration of materials caused by the reaction of materials with the surrounding environment. For metal materials, metal corrosion refers to the destruction phenomenon caused by the chemical or electrochemical action of the metal surface and surrounding media [4]. The result is a change in the properties of the metal and often damage to the function of the metal, the environment or the technical system of which they are a part.

The main forms of metal corrosion are chemical corrosion and electrochemical corrosion, and the environmental factors that play a role in corrosion mainly include the following aspects [5]:

- Medium: hydrogen ions, chloride ions and other media have a speeding effect on metal corrosion.
- Temperature: Normally, corrosion accelerates as the temperature increases.
- Flow rate: Too fast or too slow will increase the speed of corrosion.

In general, anti-corrosion design is mainly achieved through anti-corrosion coatings and corrosion inhibitors for protection. However, negative ions in the environment have a destructive effect on anti-corrosion coatings, so it cannot fundamentally eliminate corrosion and mainly plays a role in suppressing the corrosion rate. Therefore, in the design and operation of civil aircraft APU, the anti-corrosion design can only slow down the corrosion rate, and the anti-corrosion work in the whole life cycle of APU must be supplemented by inspection and measures of operation and maintenance.

3 APU Anti-Corrosion Design

The APU is connected to the atmosphere through the intake and exhaust systems and can be directly exposed to the humid air outside, as shown in Fig. 1. APU anti-corrosion design usually includes material selection considerations and subsequent storage procedures development and implementation.

There are usually several reasons for APU corrosion:

- APU corrosion causes usually exist for the following reasons:
- APU anti-corrosion design is not in place.
- The APU storage program is not properly developed.
- The APU stored program is not executed properly.
- Prolonged harsh environment.

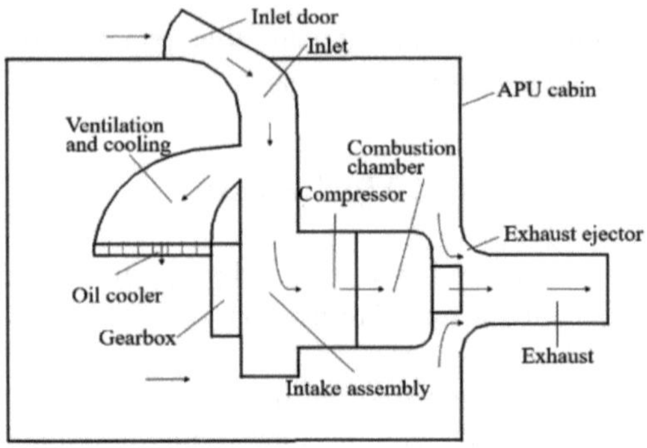

Fig. 1. APU system diagram

3.1 Anti-Corrosion Design

Usually, anti-corrosion design mainly relies on anti-corrosion coatings and the corrosion resistance characteristics of the materials themselves. At present, the APU compressor is usually made of titanium alloy, and the turbine guide is usually made of super alloy such as nickel-based alloy. Both materials have good corrosion resistance. Turbine components are coated with thermal barrier coating, which can avoid direct contact with high temperature flame, effectively reduce component temperature, and improve thermal corrosion resistance.

3.2 APU Stored Program

APU storage methods are usually divided into on-wing storage and off-wing storage. For on-wing storage, it is usually divided into APUs that can operate regularly and APUs that cannot operate regularly.

For APUs that can be operated regularly, it is usually necessary to place drying bags at the intake and exhaust ports and install protective covers as needed, depending on the humidity. Ensure that the APU is operated regularly at specified intervals.

For APUs that cannot operate regularly, it is usually necessary to install intake and exhaust protective covers, drain fuel and lubricating oil, place desiccant and humidity indicators, and check the humidity regularly. If the humidity exceeds the limit, replace the desiccant.

Usually APU off-wing storage program mainly includes the following steps:

- Drain fuel and lubricating oil.
- Place desiccant and humidity indicator.
- Install protective covers in all open parts of the engine.
- Pack the APU and place desiccant and humidity indicator in the box.
- Stored in a temperature and humidity controlled environment.

For an APU in a stored program, it is necessary to release the stored program to use or enter the next storage cycle. The normal program consists of the following steps:

- Remove all protective covers, desiccant, and humidity indicators.

- Add oil to APU.
- Visually inspect cable connections and APU component integrity.
- Low speed running APU (avoid APU stuck, full speed running cause danger).
- Full speed APU (usually running for more than 5 min).

4 Effect of Corrosion on APU

Since corrosion occurs on rotor components, if the corrosion goes deep inside the substrate, it will increase the probability of APU rotor rupture. According to TSO-C77b requirements [6], For each high-energy APU rotor, the APU must be designed to provide containment of the largest blade section and maximum kinetic energy fragments. In normal APU, rotor containment will be considered, that is, containment rings are set outside the compressor and turbine. Only when the rotor containment component fails and the rotor disk is broken due to superimposed corrosion, the non-containment rotor blasting event will occur.

Corrosion is accompanied by peeling of the coating. The coating shedding will not affect the strength of the substrate, but may damage the blade, affect the surface flow field or reduce the heat resistance of the hot end components, resulting in reduced performance and life of the APU, serious corrosion phenomenon needs to be returned to the factory for overhaul, increase the maintenance cost, and may cause the APU to fail to start or unordered shutdown.

5 The APU Does not Perform Subsequent Operations of the Stored Program as Required

Due to the usual material characteristics of APU, if it is parked in a high salt and high humidity environment for a long time, without taking appropriate protective measures, APU is prone to corrosion. Therefore, if the APU is not used for a long time, it needs to run the corresponding APU stored program to ensure a dry environment in the APU and prevent corrosion of the APU.

If the APU does not execute the stored program as required, the impact on the APU cannot be easily determined. In general, the following schemes are used to determine whether the APU can continue to be used:

- Perform APU borescope detection according to the program to observe the corrosion inside the APU.
- Manually rotate the APU and check whether there is friction or stagnation during rotation.
- Collect oil sample from oil filter element/filter cap to test oil water content and total acid value.

6 Specific Case Analysis

Several APU corrosion events occurred on the route of a certain type of APU. There is obvious yellow rust on the metal surface, the surface is uneven, and there is peeling of the paint layer. The specific corrosion examples are shown in Fig. 2.

Fig. 2. Example of APU corrosion

6.1 Physical and Chemical Analysis

The corrosion surface was sampled, and the results of energy spectrum analysis were shown in Fig. 3. The surface samples contain more oxygen, Sulphur and other elements, which is presumed to be oxidizing corrosion.

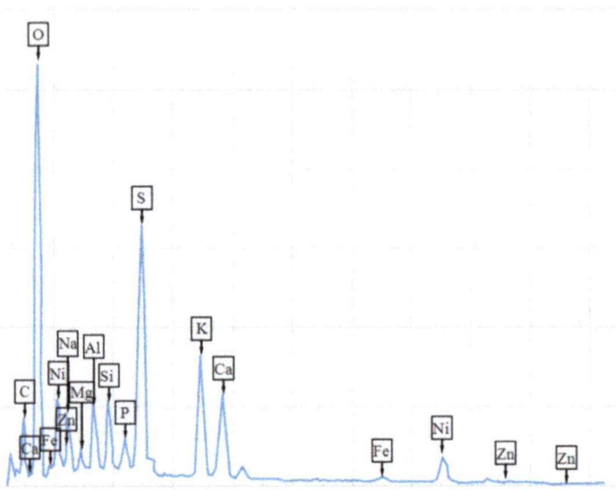

Fig. 3. Surface physical and chemical analysis

6.2 Corrosion Cause Investigation

In view of the corrosion phenomenon of APU, the borescope detection inspection of APU was carried out. The results of borescope detection show that APU corrosion mainly occurs in compressor casing and secondary turbine guide blades as shown in Table 1. A total of 21 APUs were inspected, of which 9 were found to be corroded.

For this event, fault tree analysis was adopted, as shown in Fig. 4. Among them, the APU has the same series APU, and the core anti-corrosion design is the same, but

Table 1. APU corrosion investigation

Borescope condition	Number of APU
There is a small amount of spot-like corrosion on the blade back and pipe of the secondary turbine guide.	6
Compressor casing slightly corroded.	3
No corrosion	12

the same series APU does not have corrosion phenomenon, so unreasonable material and coating design is excluded (EN1.1). The APU drain design has been verified as seaworthy, so the body drain design is ruled out as unreasonable (EN1.2). Compared with the stored program of the same series of APUs, this APU is more stringent, so the stored program problem is excluded (EN2). After investigation, all corroded APUs have experience of not performing according to storage requirements, and all have high salt and high humidity environment operation. Other APUs of the same model operating in high-salt and high-humidity environments were investigated simultaneously and stored procedures were executed as required. No corrosion occurred. Therefore, the corrosion event is strongly associated with non-execution of stored program (EN3).

To sum up, the direct cause of this corrosion phenomenon is that the APU is not maintained according to the manual requirements, which causes the APU to be exposed to humid environment for a long time. The aluminum-based wear-resistant coating on the compressor cover shell absorbs water and expands, resulting in coating corrosion and material uplift. For the thermal barrier coating of the hot-end parts, the coating will fall off due to ablation under high temperature working environment. At the same time, dust, salt and other small particles in the air/gas are attached to the secondary guide area in the middle temperature area at the back end of the hot end component. If the area is in a high humidity environment for a long time, the thermal barrier coating is prone to further electrochemical corrosion and fall off, which affects the integrity of the substrate.

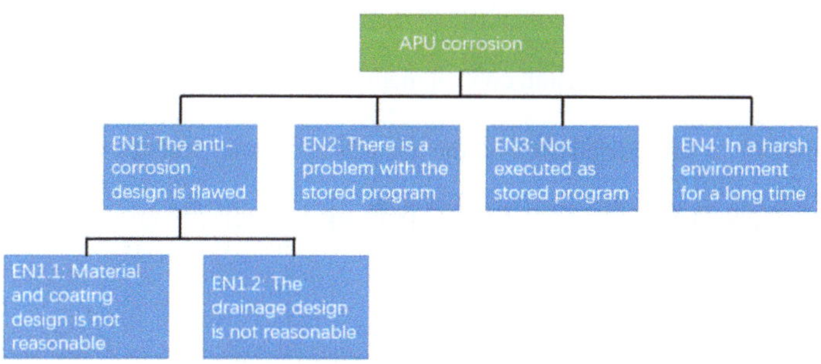

Fig. 4. APU corrosion fault tree

7 Experience and Lesson

Programmed execution of the stored program of the APU is essential to the use of the APU. The airline incident has given people enlightenment, although not as required to implement or simplify the APU stored procedures, at the time to save man hours, convenient operation. However, in the long run, failure to implement or simplify the storage program of the APU as required can easily lead to corrosion, reduce the performance and life of the APU, and in serious cases, return to the factory for overhaul, reducing the overall economy of the APU. The cost far exceeds the time saved.

APU corrosion protection is guaranteed through corrosion protection design and the development of reasonable storage procedures. Only through effective means to ensure that the stored program can be executed correctly and in time, in order to ensure that all the efforts in the early stage will not fall short because the final implementation is not in place.

8 Conclusion

APU storage program execution is not in place, combined with high temperature and humidity environment, is an important factor causing APU corrosion. APU corrosion may cause rotor disk rupture, and failure of superimposed containment components may lead to non-containment rotor blasting events. At the same time, it will also cause the performance of APU to decline and affect the life span, so the APU anti-corrosion is crucial to the life course of APU.

References

1. Ling, K.: Discuss future trends of APU from the comparison of three models APU. Popular Sci. Technol. **15**(1), 2 (2013)
2. Huo, W., Sun, H.: Corrosion and precaution for blades of aero-engine compressor in the navy air arm. Aviation Maintenace **6**, 39–41 (2002)
3. Xu, X., Ao, L., Wu, Z.: Study of corrosion in coastal zone. Modem Ind. Econ. Informationization **11**, 179–182 (2021)
4. Zhu, Y., Ye, Z.: Metal corrosion and control methods of boiler water and steam system. Chem. Eng. Equip. (8), 3 (2019)
5. Li, A., Wang, W.: A brief discussion on the hazards and protection of metal corrosion. Chem. Manage. (12), 1 (2013)
6. TSO-C77b, 'GAS TURBINE AUXILIARY POWER UNITS' (2000)

Strategy for Sustainable Electrification and Renewable Energy Innovations in South Sudan

Jacob Manyuon Deng[1(✉)], Cyrus Wabuge Wekesa[2], Khan Jean De Dieu Hakizimana[1], and Joseph Nzabahimana[3]

[1] ACE-ESD/College of Science & Technology, University of Rwanda, P.O. Box 4285 Kigali, Rwanda
makouyar@gmail.com
[2] School of Engineering, University of Eldoret, Eldoret 1125-30100, Kenya
[3] College of Education, University of Rwanda, Eastern Province, P.O. Box 55 Rwamagana, Rwanda

Abstract. Background: South Sudan is a land lock country in East - Central Africa and ranks the least electrified country in Sub-Saharan Africa. Most of its power supply comes from biomass and diesel fuel that has led to environmental and health issues such as deforestation and air pollution. Electricity prices are exorbitant and hence, unaffordable. The existing electricity grid in South Sudan's towns is in dilapidated state and is limited to the key state cities. This network is not capable of extending to most customers in towns and the rural areas. The purpose of this paper is to analyze the implementation strategy for the sustainable electrification and renewable energy innovations in South Sudan. The paper looks at renewable energy potentials with decentralized electrification focus through the Multi-tier Framework for energy access approach. The paper evaluates wind and solar potential in four selected key locations and presents results pictorially. Data was collected from South Sudan Civil Aviation, Metrological Department and through the structured interviews from Doleib Hill village. The data was analyzed using Rural Africa Load Profile Tool (RALPT), XLSTAT2023, Tabular and Google Sheets tools to display the results.

Results: The analysis shows that South Sudan has potential for wind and solar energy that could be an epitome for sustainable electrification strategy in the country. The wind potential in four locations of Juba, Malakal, Raga and Wau, with Juba coming out with 220.16 W/m^2 in April, Malakal has 373.31 W/m^2 in January and 282.81 W/m^2 in December while Wau and Raga have relatively low power densities of 136.85 W/m^2 and 133.72 W/m^2 respectively signify favorable conditions for renewable energy projects. Solar radiation levels that range from 17.19 MJ/m^2/day in July to 21.39 MJ/m^2/day in January in Juba which consistently falls within the "High Solar Radiation" category although with reduced solar radiation. Malakal on the other hand, ranges from 16.83 MJ/m^2/day in July to 21.39 MJ/m^2/day in February. Raga has high solar radiation from 17.14 MJ/m^2/day in July to 21.62 MJ/m^2/day in February ensures consistent solar energy production and system reliability. The data analysis in Doleib Hill shows 65% of respondents falling into Tier 0, showing reliance on wood, charcoal, and animal waste, with no access to modern electricity. Only 1% of these respondents

T. Zielinska (Ed.): PEAE 2025, LNNS 1679, pp. 179–193, 2026.
https://doi.org/10.1007/978-3-032-08221-3_17

use generators, while 9% have access to small solar systems. Tiers 1 and 2 give 10% of households and private sector respondents that use lighting and phone charging. 7% on biomass, 6% on generators and 7% on solar. Tiers 3 to 5 give 25% of NGOs and Public organizations more reliable energy. 4% on biomass, 7% on generators and 6% on the solar system. The results present energy potential for decentralized electrification.

Conclusion: Power supply in South Sudan is inadequate and expensive. It is limited to a few state capital cities. Rural areas have no electricity due to lack of power generation and associated grid systems. The locations of Juba, Malakal, Raga and Wau have high wind and solar potential throughout the year. A Multi-Tier Framework for Energy Access is a viable electrification strategy through mapping and harnessing the renewable energy sources in the country. This strategy presents an in-depth assessment of accessibility rates by means of non-binary approach while adopting off grids connectivity solutions such as microgrids, solar home system, solar lantern etc. Doleib Hill model is a candidate for reciprocation around the country. With this strategy, the country could be covered with electricity at a faster rate, with the development of renewable energy sources.

Keywords: South Sudan · Sustainable Electrification · Multi-Tier Framework · Renewable Energy · Decentralized Electrification

1 Introduction

South Sudan covers a land area of 644,329km^2 and has a population of around 13 million [1]. It became an independent nation on July 9, 2011, making it number 54 in Africa and 193rd to join the United Nations (UN). The Country joined the East African Community (EAC), in September 2016, to become the 7th Member state. South Sudan shares its borders with Sudan to the north, Ethiopia to the east, the Central African Republic to the west, and Kenya, the Democratic Republic of Congo, and Uganda to the south. It is a landlocked country and it falls between latitudes 3° and 13°N and longitudes 24° and 36°E [2].

Access to reliable and sustainable energy is a pivot in driving economies. Sub-Saharan Africa (SSA) continues to face significant challenges with electricity access. Figure 1 presents a global population without access to electricity with Sub-Saharan Africa accounting for 40%, South Asia, 25%, East Asia and Pacific, 15%, West Asia and North Africa, 10% and Latin America and Caribbean with 10%. This signifies that over 80% of the world's population without electricity is underdeveloped. As part of SSA, South Sudan situation is dire, as the country grapples with alarmingly low levels of electricity access, with only 7.4% of the population having access to electricity [3–5].

Energy is a crucial factor in driving sustainable development as it powers various aspects of modern infrastructure. Energy ensures economic growth while minimizing environmental impact as it sustains transportation, industry and agriculture. Access to clean energy is key for development and technological advancement [6]. The population of South Sudan continues to grow from 2005 to 2024. Taking the year 2010 and 2011, there was an increase of 1.05% and from 2020 to 2021, the population increased with 1.1%. In 2024, there was an increase of 1.7% compared to 1.47% increase in 2023

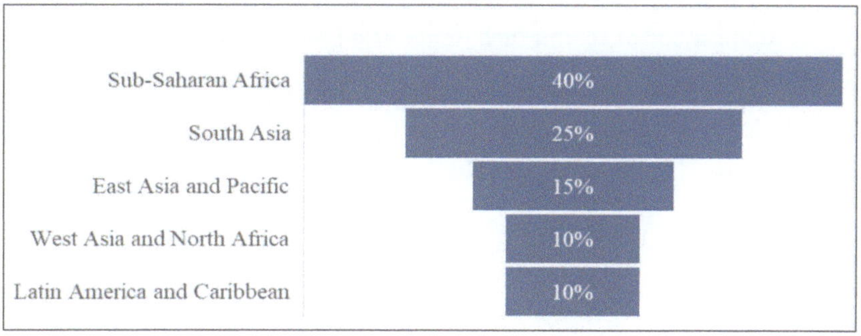

Fig. 1. Global populations without access to electricity 2024 [3–5]

(Fig. 2 gives more details) [8, 9]. Even though the population has grown, the country still struggles to achieve access to reliable power supply [10–13].

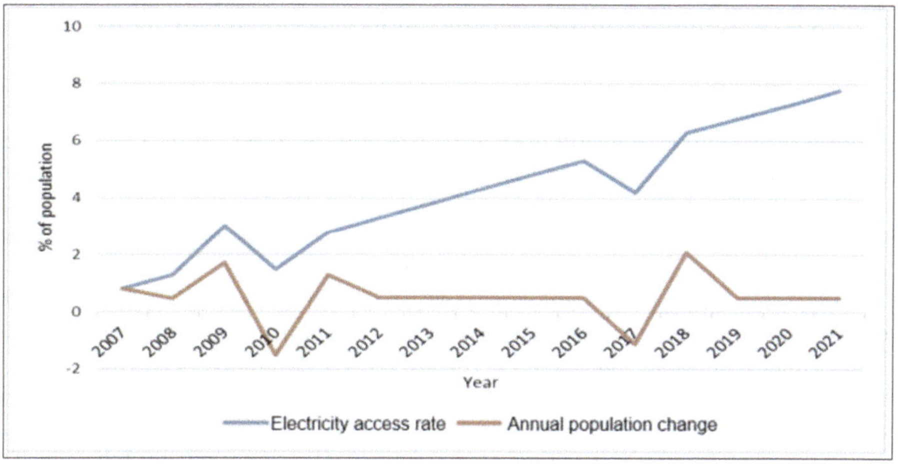

Fig. 2. Electricity access rate and annual population change [8, 9]

Greater portion of the generation mix in South Sudan is dominated by diesel, which is quite expensive, unreliable, and inaccessible. Though South Sudan is an oil producing country, it still lacks domestic refining capacity and hence depends on imported fuel. The total installed generation capacity stands at approximately 141 MW compared to demand estimates of over 300 MW [12]. Of this capacity, more are either non-operational or serve the country's oil fields and are not accessible to the public. Estimate indicates that less than 35 MW of power is currently serving the public. Power supply is limited, and is managed through load shedding and power outages [13], rendering the power sector ineffective to manage the power supply. Figure 3 gives an example of the capital city of Juba, which reveals nonexistence of renewable energy sources.

The country may not develop with 83% of its population lacking access to electricity [15, 16]. Traditional/conventional electricity access strategy has been Grid Extension,

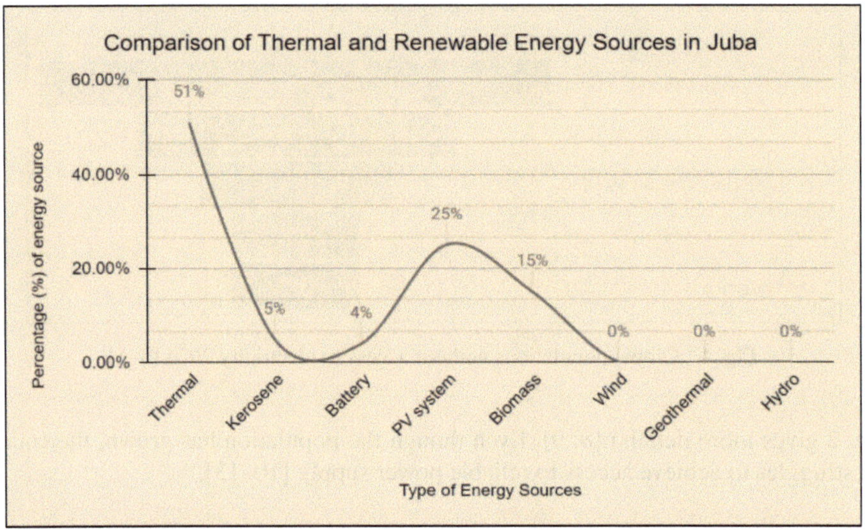

Fig. 3. Thermal and Renewable Energy Sources in Juba (MED, 2022), [14]

which is nonexistence in South Sudan [17]. However, this strategy has challenges due to low power consumption in the rural areas. Grid extension, as an electrification strategy, is a burden to utilities, with strong and diverse revenue implications. The per capita energy consumption for South Sudan is as poor as about 41kWh per person compared to 180 kWh in Sub-Saharan Africa [18]. This indicates poor electricity situation in South Sudan which may not support economic growth.

South Sudan energy needs are dependent on fossil fuel and biomass from animal and agricultural produce. Diesel is about 90% and it is utilized by both individuals and utilities. Other renewable sources of energy are not yet developed [19]. About 70% of the populace depends on charcoal, wood-fuel, animal dung and crop residue, to perform cooking and heating [20]. According to the Ministry of Energy and Dams, 100% of power is diesel and majorly an estimate with only 33MW operational in Juba city. The estimated demand is diesel fired and is projected for cities that are attractive industries [21].

In 2015, investment in oil and Solar Photo Voltaic (PV) grew after the 2013 fighting and dropped in 2016 when fighting recurred in the capital city. From 2017 to 2020, displaced persons and those in rural areas had moved to the state cities and therefore oil had increased and remained constant [22]. Figure 4 explains the pictorial chart of the values of Urban, Rural, and total percentage rate over the years from 2010 to 2020 according to the SDG 7 tracking. Access is higher in urban cities compared to rural areas. The, total access to electricity had increased considerably due to rural-urban movements [18]. The total access rate from 2010 to 2020 is affected by the contribution of low access in the rural areas.

Figure 5 shows that demand is high and has surpassed the generation. Besides, agriculture and industries are growing despite unreliability of power supply. Diesel that had risen linearly in 2015 and 2020 may not meet the fast-growing demand in the country,

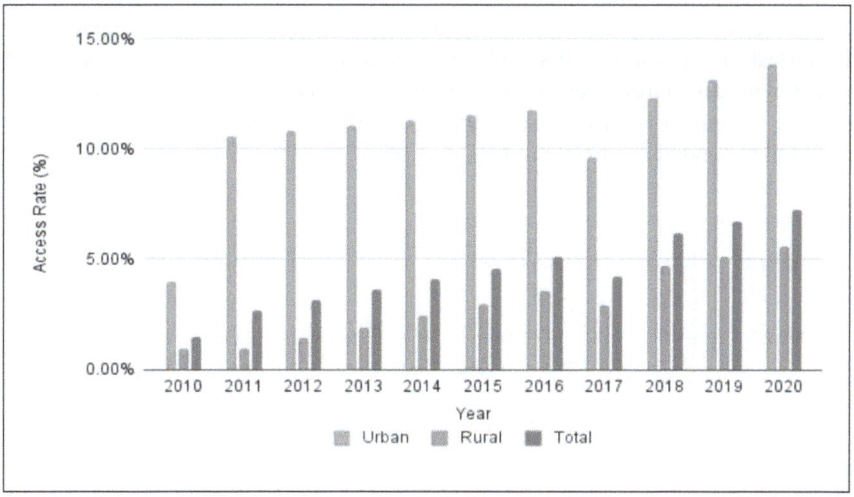

Fig. 4. South Sudan Electricity Generation Capacity. Source IEA Database [23].

hence an impediment to major investment due to excessive cost. Small scale industries such as water factories started growing from 2017 to 2020. The country is endowed with small-scale agricultural resources while commercial activities have increased the demand over the years.

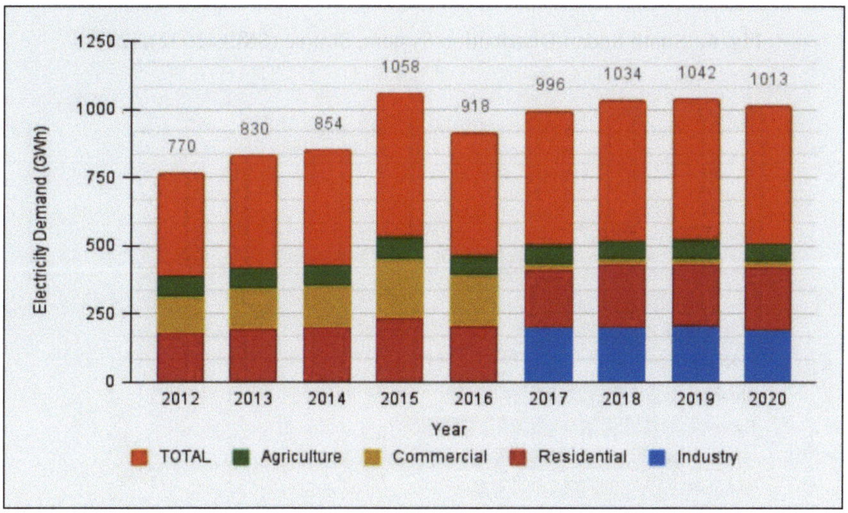

Fig. 5. South Sudan Electricity Demand. Source IEA database [23].

Cities such as Malakal, Wau, Rumbek and Yambio had 11kV distribution networks. This network was vandalized during the 2013 war. Figure 6 shows the concentration of distribution network in the state cities [24]. Looking at the distance in Square Kilometers for the state to cover with the network, it is not possible to achieve adequate coverage in the shortest time possible. This analysis presents a clear picture that many customers in those towns can hardly get power through these isolated grids. There is not any distribution system in rural areas and the distribution systems in the state cities have been vandalized and are not functioning.

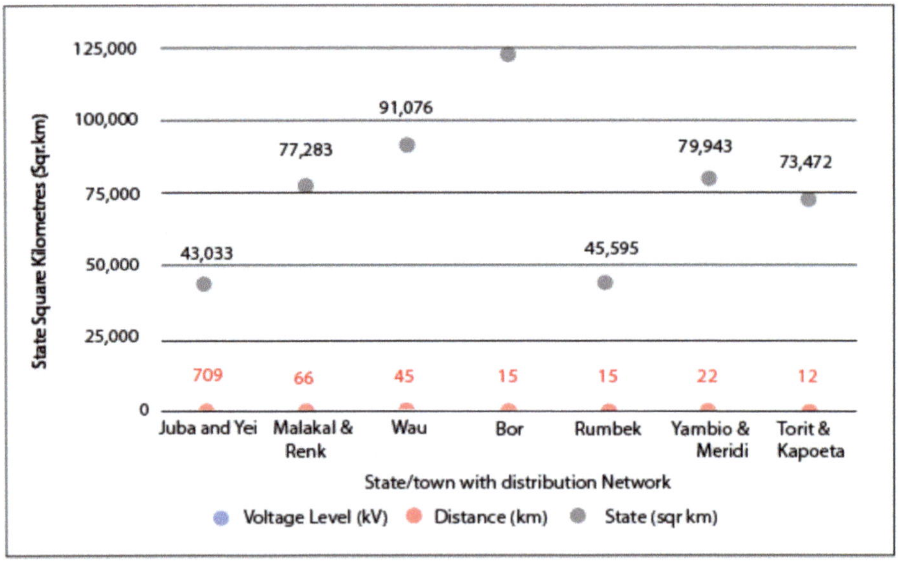

Fig. 6. South Sudan Distribution System, Source (SSEC 2022), [24]

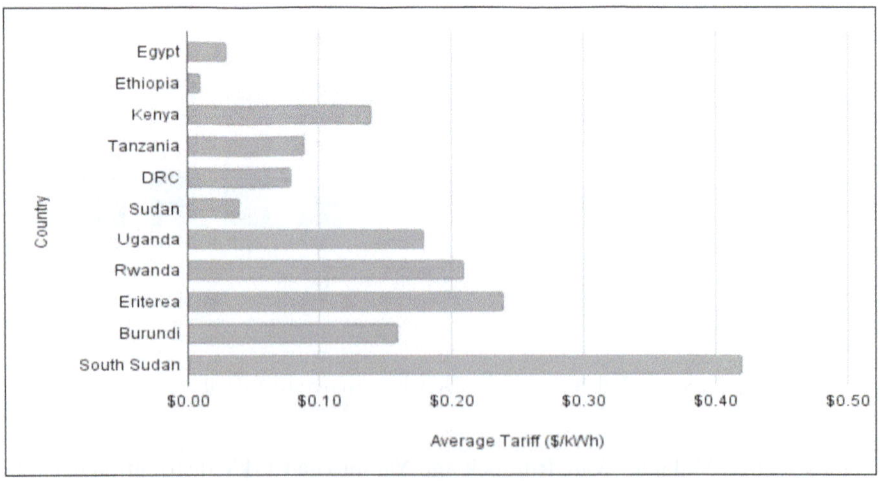

Fig. 7. Comparison of Electricity Tariff Rates in Sub-Saharan African Countries, [24].

Electricity prices in South Sudan are not affordable. Figure 7 shows the comparison of electricity tariff rates among the SSA countries. South Sudan recorded the highest tariff rate due to the type of the energy sources in use. High cost is a major barrier which translates into affordability issues.

This paper analyzes sustainable electrification implementation strategy for renewable energy innovations in South Sudan. For a proper organization of this paper, we start with a brief introduction of South Sudan and electricity supply situation in Sect. 1. Section 2 presents the analysis of sustainable electrification strategies and renewable energy innovations, Sect. 3 discusses the methodology used to collect the data and the analysis of the results. Section 4 concludes the discussion and gives recommendations for policy decisions.

2 Analysis of Sustainable Electrification Strategies and Renewable Energy Innovations in South Sudan

With a heavy reliance on diesel, the power supply in South Sudan is not consistent and is characterized with outages and shutdowns. Customers are forced to use owned generators and rooftop solar arrays in the cities. Charcoal, wood, grass, cow dung, and agricultural residues, are the primary source of energy consumption in South Sudan's rural areas [25, 26]. The high dependence on biomass for energy in South Sudan poses significant environmental and health challenges due to deforestation, air pollution, and respiratory diseases. One major consequence is the alarming rate of forest loss, with the country losing approximately 1.5% to 2% of its forests on annual basis [27].

Renewable Energy Potential

There is an urgent need to explore alternative and sustainable energy sources to mitigate the energy crisis in South Sudan and diversify its energy sources and improve access to electricity, given that South Sudan has great potential for wind and solar energy. It is, therefore, an opportunity for the country to develop its energy sources and reduce dependency on fossil fuels [25, 26]. Investing in wind and solar energy infrastructure can help South Sudan to improve the environment and public health and meet its growing energy demand with a reliable and affordable rate. It is imperative to evaluate the wind power density and solar irradiance of individual sites for renewable energy initiatives.

The investments in power generation in South Sudan had concentrated in Juba. There is no power supply in other major cities of Malakal and Wau since power infrastructure was destroyed during the war in 2013 and 2016. The current functioning power systems in Juba is 100% diesel [28]. Figure 8 presents available renewable energy sources in Juba. The PV which accounts for 25% is utilized by individuals who can afford the cost, Biomass which is also used by low level class through charcoal, animal dungs and farm residues accounts for 15% while other Renewable Energy System (RES) are not exploited. This is giving way to the country to venture into a large-scale investment in renewable energy fields.

Decentralized Electrification

Expanding South Sudan's national grid to achieve universal access is not possible or cost-effective in the short term due to the current state of the country's energy infrastructure.

High costs, expansive geographical area, and low population density in rural areas hinder provision of electricity. However, in this context, decentralized electrification proves to be a practical alternative. It involves the utilization of localized energy solutions such as micro-grids, home systems, and mini-grids [29–31]. Decentralized electrification offers numerous advantages. It enables the customization of energy solutions to suit the unique requirements of individual communities. This approach can be swiftly implemented at a lower cost than grid expansion, making it suitable especially for remote areas. Furthermore, decentralized systems can harness renewable energy sources, offering a sustainable and environmentally friendly substitute for fossil fuels. This is not only aiding the reduction of the country's carbon footprint but also reduces reliance on expensive and unreliable fuel imports [32, 33].

The Multi-Tier Framework (MTF) for energy access enhances this process by offering systematic evaluation of energy access. Unlike the simple binary of "connected" versus "not connected," the MTF assesses energy access across multiple tiers, considering aspects such as capacity, quality, reliability, safety and affordability. Each tier represents a different level of energy service, from Tier 0 (no access) to Tier 5 (full access). This multi-dimensional approach is especially valuable in regions with diverse energy needs, allowing for a detailed understanding of energy access and enabling tailored solutions. For instance, while some areas may only need basic services like lighting and phone charging (Tier 1), others may require more advanced services such as refrigeration and productive energy uses (Tier 4 or 5) (Brown & Garcia, 2023). This structured analysis supports the development of more effective and context-specific renewable energy solutions [34–36] (Table 1).

Table 1. Multi-Tier Framework, [37]

MTF Tier	Service Capacity	Example Services	Daily Availability	Evening Availability	Reliability	Quality	Affordability
Tier 0	No electricity	None	None	None	Not applicable	Not applicable	Not applicable
Tier 1	Minimum 3W (typically solar lanterns)	Task lighting, phone charging	Minimum 4 h	Minimum 1 h	Not required	Basic	Cost of electricity <5% of household income
Tier 2	Minimum 50W	General lighting, television, fan	Minimum 4 h	Minimum 2 h	Not required	Basic	Cost of electricity <5% of household income
Tier 3	Minimum 200W	Multiple lights, television, refrigerator, basic appliances	Minimum 8 h	Minimum 3 h	<14 disruptions per week	Stable	Cost of electricity <5% of household income
Tier 4	Minimum 800W	High-power appliances (e.g., washing machine, iron)	Minimum 16 h	Minimum 4 h	<3 disruptions per week	Good	Cost of electricity <5% of household income
Tier 5	Above 2000W	Full household electrification, including AC	24 hours	Minimum 4 h	<3 disruptions per week	Good	Cost of electricity <5% of household income

3 Methodology

This paper evaluates the wind and solar potential in four key locations of Juba, Malakal, Raga, and Wau. These cities were chosen due to their strategic geographic positioning and economy [18]. The paper utilizes 40 years of historical Wind and solar data from

1974 to 2014, from the South Sudan Civil Aviation. The wind data was recorded at a height of 10 m. The dataset helps analyze yearly and monthly variations of wind and solar energy potential. The wind power density is computed using the following formula [39, 40]. Table 2 shows the classification of standard wind power density suitable for turbines.

Table 2. Wind power density classes at 10 m.a.g.l

WPD Class	WPD (W/m^2)	Suitability
Class 1	<50	Poor
Class 2	50–100	Marginal
Class 3	100–150	Fair
Class 4	150–200	Good
Class 5	200–250	Very Good
Class 6	250–300	Excellent
Class 7	>300	Outstanding

$$WPD = \frac{1}{2}\rho v^3 \qquad (1)$$

- WPD is the wind power density (in watts per square meter, W/m^2),
- ρ is the air density (in kilograms per cubic meter, kg/m^3), in this study the standard air density of 1.225 kg/m^3 has been used.
- And v is the wind speed is (in m/s).

The Angstrom-Prescott model which relies on a linear correlation between the ratio of the actual sunshine duration to the maximum possible sunshine duration and the ratio of the actual solar radiation to extra-terrestrial radiation was used to determine the solar radiation. This model is expressed as [41]. Table 3 presents the standard classification of solar radiation around the world.

$$H(\mathrm{MJ/m/day}) = H_0\left(a + b\frac{S}{S_0}\right) \qquad (2)$$

where:

- a = 0.22 and b = 0.54
- S is the mean daily number of sunshine duration (h)
- H_0 is the extra-terrestrial Radiation
- S_0 is the maximum possible daily sunshine duration (h)

Table 3. Solar radiation Classification ($MJ/m^2/day$) [52]

Classification	Solar Radiation ($MJ/m^2/day$)	Description
Very High Solar Radiation	Above 25	Ideal for concentrated solar power (CSP) and large-scale photovoltaic (PV) systems.
High Solar Radiation	20–25	Suitable for various solar technologies, including PV and solar thermal systems.
Moderate Solar Radiation	15–20	Adequate for most solar applications, including PV and solar water heating.
Low Solar Radiation	10–15	May limit the efficiency of solar power systems, suitable for small-scale solar applications.
Very Low Solar Radiation	Below 10	Limited solar energy potential; suitable for niche applications like solar-powered calculators.

It is also based on quantitative and qualitative methods of data collection to assess decentralized electrification strategy in Doleib Hill Village. Rural Africa Load Profile Tool (RALPT) is applied to analyses the demand while, multi-Tier framework is used to evaluate energy access levels and potential impacts of renewable technologies. The results contribute to developing a model for decentralized electrification, offering a potential solution for bridging the energy access gap in similar communities across the country. Cochran's formula for a finite population is used to determine sample size as shown below:

$$n = \frac{N.Z^2.p.(1-p)}{e^2.(N-1) + Z^2.p.(1-p)} \tag{3}$$

where:

- n = required sample size
- N = total population size
- Z = Z-value (e.g., 1.96 for 95% confidence level)
- p = estimated proportion of the population (assumed to be 0.5 for maximum variability)
- e = margin of error (e.g., 0.05 for a 5% margin)

Tools and Analysis

The paper analyses historical daily wind and solar data collected over a period of 40-years (1974–2014) from four meteorological stations in South Sudan. It further highlights significant reliance on biomass and diesel generators due to the absence of a centralized grid. The paper also assesses the load demand through structured interviews and categorizes the levels of consumption with multi-tier analysis, defining Tiers; from 0 to 5 to ascertain sustainable electrification strategy on Doleib Hill village in South Sudan with focus on wind power density and solar radiation. XLSTAT 2023, Tubular, Google Sheet tools and Angstrom-Prescott model were used.

Results

The analysis shows that South Sudan has potential for wind and solar energy that could be an epitome for sustainable electrification strategy in the country. Figure 9 gives the wind potential in four locations of Juba, Malakal, Raga and Wau, with Juba coming out with 220.16 W/m^2 in April, Malakal has 373.31 W/m^2 in January and 282.81 W/m^2 in December while Wau and Raga have relatively low power densities of 136.85 W/m^2 and 133.72 W/m^2 respectively, making them categorized as "Marginal" to "Fair" levels.

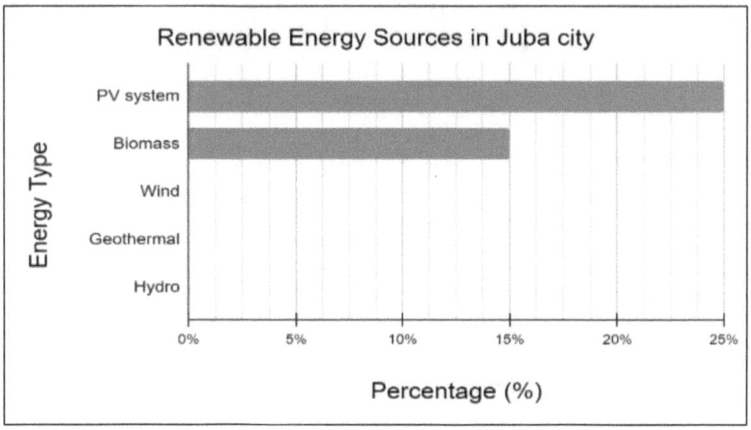

Fig. 8. Renewable energy Sources in Juba, (SSEC, 2022), [28]

Figure 10 presents solar radiation levels that ranges from 17.19 MJ/m^2/day in July to 21.39 MJ/m^2/day in January in Juba which consistently falls within the "High Solar Radiation" category although with reduced solar radiation. Malakal on the other hand, ranges from 16.83 MJ/m^2/day in July to 21.39 MJ/m^2/day in February. Raga has high solar radiation from 17.14 MJ/m^2/day in July to 21.62 MJ/m^2/day in February ensures consistent solar energy production and system reliability.

The data analysis shows 65% of respondents falling into Tier 0, showing reliance on wood, charcoal, and animal waste, with no access to modern electricity. Only 1% of these respondents use generators, while 9% have access to small solar systems. Tiers 1 and 2 give 10% of households and private sector respondents that use lighting and phone charging. 7% on biomass, 6% on generators and 7% on solar. Tiers 3 to 5 give 25% of NGOs and Public organizations more reliable energy. 4% on biomass, 7% on generators and 6% on the solar system. Figure 11 shows aggregate multi-Tier respondent's distribution.

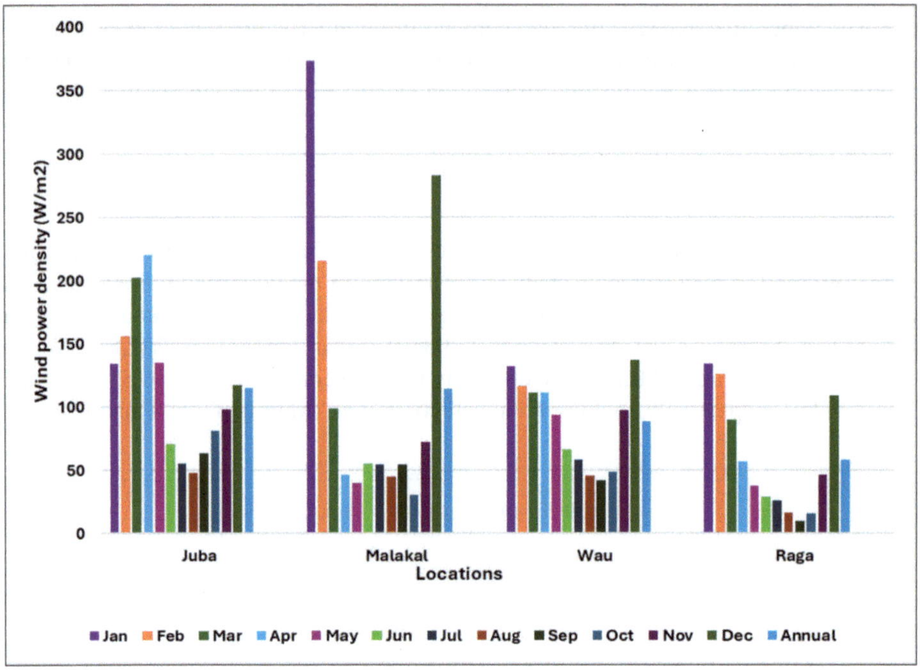

Fig. 9. Monthly wind power density

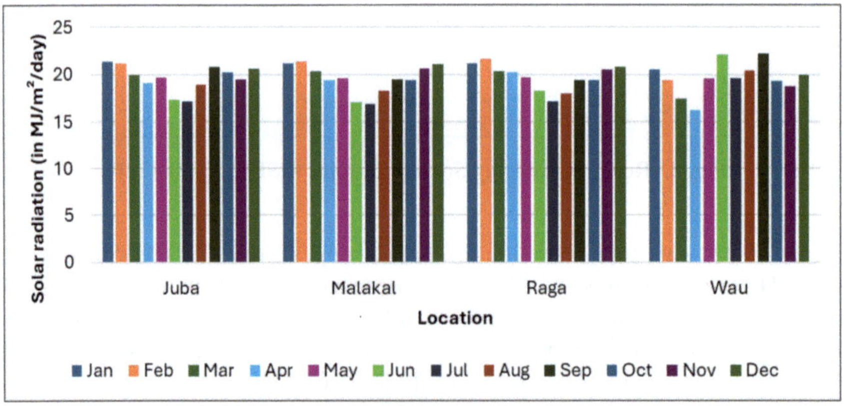

Fig. 10. Monthly solar radiation

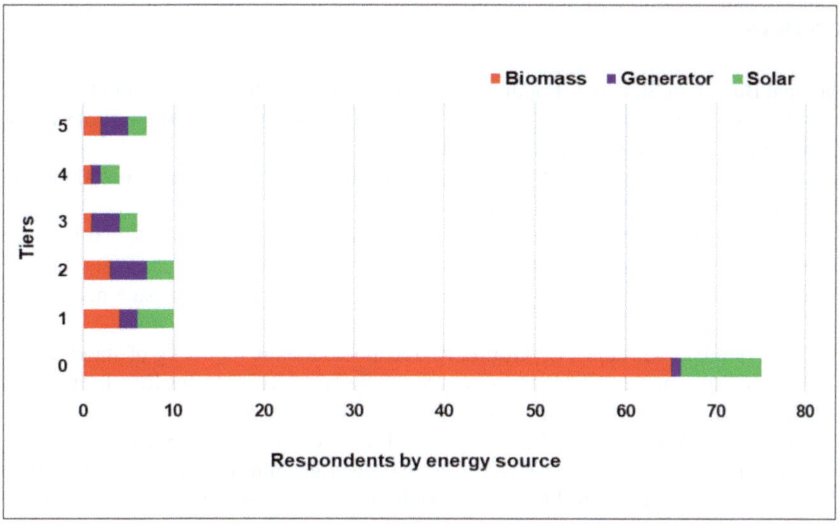

Fig. 11. Aggregate MTF respondent's distribution

4 Conclusions and Recommendations

4.1 Conclusions

Power supply in South Sudan is inadequate and expensive. It is limited to few state capital cities. Rural areas have no electricity due to lack of power generation and associated grid systems. The locations of Juba, Malakal, Raga and Wau have high wind and solar potentials through the year. A Multi-Tier Framework for Energy Access is a viable electrification strategy through mapping and harnessing the renewable energy sources in the country. This strategy, presents an in-depth assessment of accessibility rates by means of non-binary approach while adopting off grids connectivity solutions such as microgrids, solar home system, solar lantern etc. Doleib Hill model is a candidate for reciprocation around the country. With this strategy, the country could be covered with electricity at a faster rate, with development of renewable energy sources.

4.2 Recommendations

1. Implementation of hybrid systems with focus on off-grid communities across South Sudan.
2. Provision of financial incentives to renewable energy investors to facilitate their businesses
3. Integration of decentralized renewable energy systems into the national energy plans to present better understanding and ownership
4. Roll out Rural electrification programs to enhance wider accessibility through the Multi-Tier Framework (MTF) for energy access
5. Establishment of electricity regulatory framework to protect investors in the energy sector.

References

1. National Bureau of Statistics, Population projections for South Sudan by County 2015–2020, p. 24 (2015). http://www.ssnbss.org/sites/default/files/2016-08/population_projections_for_south_sudan_by_county_by_sex_2008_2015.pdf
2. National Bureau of Statistics (NBS), National Baseline Household Survey 2009 Report for South Sudan, pp. 1–175, (2012). http://www.ssnbs.org
3. Rivki, M., Bachtiar, A.M., Informatika, T., Teknik, F., Indonesia, U.K.: WIND Energy Explained: Theory, Design and Application, no. 112. John Wiley & Sons Ltd. (2009)
4. Access to electricity improves slightly in 2023, but still far from the pace needed to meet SDG7 – Analysis – IEA
5. Ayik, A., Ijumba, N., Kabiri, C., Goffin, P.: Techno-economic modeling of stand-alone solar photovoltaic systems: a case scenario from South Sudan. SAIEE Africa Res. J. **115**(2), 55–68 (2024). https://doi.org/10.23919/saiee.2024.10520213
6. Elminshawy, N.A.S., Diab, S., Yassen, Y.E.S., Elbaksawi, O.: An energy-economic analysis of a hybrid PV/wind/battery energy-driven hydrogen generation system in rural regions of Egypt. J. Energy Storage **80**, 110256 (2024). https://doi.org/10.1016/J.EST.2023.110256
7. South Sudan Population – Worldometer (2024)
8. Population growth (annual %) - South Sudan | Data
9. Ayodele, T.R., Jimoh, A.A., Munda, J.L., Agee, J.T.: A statistical analysis of wind distribution and wind power potential in the coastal region of South Africa. Int. J. Green Energy **10**(8), 814–834 (2013). https://doi.org/10.1080/15435075.2012.727112
10. Patankar, N., de Queiroz, A.R., DeCarolis, J.F., Bazilian, M.D., Chattopadhyay, D.: Building conflict uncertainty into electricity planning: a South Sudan case study. Energy Sustain. Dev. **49**, 53–64 (2019). https://doi.org/10.1016/J.ESD.2019.01.003
11. Idriss, A.I., Ahmed, R.A., Atteyeh, H.A., Mohamed, O.A., Ramadan, H.S.M.: Techno-economic potential of wind-based green hydrogen production in djibouti: literature review and case studies. Energies **16**(16), 6055 (2023). https://doi.org/10.3390/EN16166055
12. African Development Fund Juba Power Distribution System Rehabilitation and Expansion Project, no (2013)
13. African Developement Bank, South Sudan : An Infrastructure Action Plan South Sudan. A Program for Sustained Economic Growth, pp. 1–321 (2013)
14. Omar, A.: Energy, environment and sustainable development in Sudan. Exergy An Int. J. **2**(1), 1–132 (2011)
15. Nile Basin Initiative Nile Equatorial Lakes Subsidiary Action Plan Program H3469553, Hydropower Expansion Plan and Regional Integration Plan of South Sudan into Regional Electricity Grid - Draft Report, (2014)
16. Environmental, S., Report, F.: Energy Sector technical Assistance
17. Energy, R.: Provision of Electricity and Rural Energy in South Sudan (2011)
18. Bank, W.: South Sudan electricity sector diagnostics. New Polit. Econ. **11**(4), 571–581 (2021). https://doi.org/10.1080/13563460600991028
19. R. S. Assessment and G. A. Report, "The Republic of South Sudan Sustainable Energy for All Rapid Situation Assessment and Gap Analysis Report (2013)
20. Tutmamer, N., Anai, J.G.: Transitioning to Renewable Energy: An Analysis of Energy Situation Special Report || Special Report ||, pp. 1–29 (2018)
21. Martin, P., List, D., Issued, D.: Bedden Hydroelectric Project Construction Planning Report For: Republic of Sudan Dam Implementation Unit (2010)
22. Ministry of Energy & Dams, Section Southern Sudan Electric Sector Policy. Report **1**(1), 0–12 (2014)

23. International Renewable Energy Agency, Renewable Energy Statistics 2022 – IRENA (2022). www.irena.org
24. Pathways To Electricity Access Expansion In South Sudan : Off-Grid And Mini-Grid Market AssessmenT Pathways to Electricity Access Expansion in South
25. Bilali, B.: Energy security in South Sudan: potentials and challenges. Turkish J. Polit. Sci. **3**(1), 1–16 (2020)
26. South Sudan - Countries & Regions - IEA
27. World Development Indicators | The World Bank
28. FabLab Nairobi, "University of Nairobi – FabLab 2017, p. 2 (2017). http://fablab.uonbi.ac.ke/
29. Bhattacharyya, S.C., Palit, D.: Mini-grid based off-grid electrification to enhance electricity access in developing countries: what policies may be required? Energy Policy **94**, 166–178 (2016). https://doi.org/10.1016/J.ENPOL.2016.04.010
30. Pearce, V.: Current and Future Trends of Electricity Service in Current and Future Trends of Electricity Service in South Sudan (2024). https://doi.org/10.1088/1742-6596/2777/1/012001
31. Peters, R., Berlekamp, J., Kabiri, C., Kaplin, B.A., Tockner, K., Zarfl, C.: Sustainable pathways towards universal renewable electricity access in Africa. Nat. Rev. Earth Environ. **5**(2) (2024). https://doi.org/10.1038/s43017-023-00501-1
32. Domegni, K.M.S., Azouma, Y.O.: Productive uses of energy: a solution for promoting energy justice in rural areas in West Africa. Renew. Sustain. Energy Rev. **160**, 112298 (2022). https://doi.org/10.1016/J.RSER.2022.112298
33. International Energy Agency(IEA), World Energy Outlook 2023 (2023)
34. D. R. Hussain, E.P., Samad, A., Koo, B.B.: BANGLADESH, Beyond connections: energy access redefined (2015)
35. Gonçalves, D.N.S., et al.: A multitier approach to estimating the energy efficiency of urban passenger mobility. Sustain. **12**(24) (2020). https://doi.org/10.3390/su122410263
36. Lahnaoui, A., Venghaus, S., Kuckshinrichs, W.: Assessing the drivers of energy supply and demand in Sub-Saharan Africa. Energy Strateg. Rev. **54**, 101483 (2024). https://doi.org/10.1016/J.ESR.2024.101483
37. Viani, L.O.: From the bottom up **93**(9) (2003). https://doi.org/10.7146/chku.v7i1.138103
38. International Renewable Energy Agency (IRENA), Renewable Energy Potential in South Sudan (2024)
39. Daoudi, M., et al.: Techno-economic feasibility of two onshore wind farms located in the desert area: a case study. Energy Sour. Part A Recover. Util. Environ. Eff. **45**(3), 7544–7559 (2023). https://doi.org/10.1080/15567036.2023.2222679
40. Idriss, A.I., Ahmed, R.A., Omar, A.I., Atteyeh, H.A., Ibrahim, M.H., Akinci, T.C.: Wind and solar energy potential in Herkalou and Lake Assal locations, Djibouti. Int. J. Power Electron. Drive Syst. **14**(1), 461–470 (2023). https://doi.org/10.11591/ijpeds.v14.i1.pp461-470
41. Vernet, A., Fabregat, A.: Evaluation of empirical daily solar radiation models for the Northeast Coast of the Iberian Peninsula. Energies **16**(6) (2023). https://doi.org/10.3390/en16062560

Author Index

T. Zielinska (Ed.): PEAE 2025, LNNS 1679, pp. 195–196, 2026.
https://doi.org/10.1007/978-3-032-08221-3

MIX
Papier aus verantwortungsvollen Quellen
Paper from responsible sources
FSC® C105338

If you have any concerns about our products,
you can contact us on
ProductSafety@springernature.com

In case Publisher is established outside the EU,
the EU authorized representative is:
Springer Nature Customer Service Center GmbH
Europaplatz 3, 69115 Heidelberg, Germany

Printed by Libri Plureos GmbH
in Hamburg, Germany